Preface

This book is based on my previous book: Tensor Calculus Made Simple, where the development of tensor calculus concepts and techniques are continued at a higher level. In the present book, we continue the discussion of the main topics of the subject at a more advanced level expanding, when necessary, some topics and developing further concepts and techniques. The purpose of the present book is to solidify, generalize, fill the gaps and make more rigorous what have been presented in the previous book.

Unlike the previous book which is largely based on a Cartesian approach, the formulation in the present book is largely based on assuming an underlying general coordinate system although some example sections are still based on a Cartesian approach for the sake of simplicity and clarity. The reader will be notified about the underlying system in the given formulation. We also provide a sample of formal proofs to familiarize the reader with the tensor techniques. However, due to the preset objectives and the intended size of the book, we do not offer comprehensive proofs and complete theoretical foundations for the provided materials although we generally try to justify many of the given formulations descriptively or by interlinking to related formulations or by similar pedagogical techniques. This may be seen as a more friendly method for constructing and establishing the abstract concepts and techniques of tensor calculus.

The book is furnished with an index in the end of the book as well as rather detailed sets of exercises in the end of each chapter to provide useful revision and practice. To facilitate linking related concepts and sections, and hence ensure better understanding of the given materials, cross referencing, which is hyperlinked for the ebook users, is used extensively throughout the book. The book also contains a number of graphic illustrations to help the readers to visualize the ideas and understand the subtle concepts.

The book can be used as a text for an introductory or an intermediate level course on tensor calculus. The familiarity with the materials presented in the previous book will be an advantage although it is not necessary for someone with a reasonable mathematical background. Moreover, the main materials of the previous book are absorbed within the structure of the present book for the sake of completeness and to make the book rather self-contained considering the predetermined objectives. I hope I achieved these goals.
Taha Sochi
London, August 2017

Contents

Preface	1
Table of Contents	2
Nomenclature	6
1 Preliminaries	**9**
1.1 General Conventions and Notations	9
1.2 General Background about Tensors	12
1.3 Exercises and Revision	19
2 Spaces, Coordinate Systems and Transformations	**22**
2.1 Spaces	22
2.2 Coordinate Systems	24
2.2.1 Rectilinear and Curvilinear Coordinate Systems	25
2.2.2 Orthogonal Coordinate Systems	26
2.2.3 Homogeneous Coordinate Systems	27
2.3 Transformations	29
2.3.1 Proper and Improper Transformations	32
2.3.2 Active and Passive Transformations	33
2.3.3 Orthogonal Transformations	33
2.3.4 Linear and Nonlinear Transformations	34
2.4 Coordinate Curves and Coordinate Surfaces	35
2.5 Scale Factors	36
2.6 Basis Vectors and Their Relation to the Metric and Jacobian	39
2.7 Relationship between Space, Coordinates and Metric	44
2.8 Exercises and Revision	46
3 Tensors	**50**
3.1 Tensor Types	50
3.1.1 Covariant and Contravariant Tensors	50
3.1.2 True and Pseudo Tensors	55
3.1.3 Absolute and Relative Tensors	57
3.1.4 Isotropic and Anisotropic Tensors	58
3.1.5 Symmetric and Anti-symmetric Tensors	59
3.1.6 General and Affine Tensors	62
3.2 Tensor Operations	62
3.2.1 Addition and Subtraction	63
3.2.2 Multiplication of Tensor by Scalar	64
3.2.3 Tensor Multiplication	64

	3.2.4 Contraction	66
	3.2.5 Inner Product	66
	3.2.6 Permutation	68
	3.2.7 Tensor Test and Quotient Rule	69
3.3	Tensor Representations	69
3.4	Exercises and Revision	71

4 Special Tensors — 76

- 4.1 Kronecker delta Tensor … 76
- 4.2 Permutation Tensor … 77
- 4.3 Identities Involving Kronecker or/and Permutation Tensors … 82
  - 4.3.1 Identities Involving Kronecker delta Tensor … 82
  - 4.3.2 Identities Involving Permutation Tensor … 83
  - 4.3.3 Identities Involving Kronecker and Permutation Tensors … 84
- 4.4 Generalized Kronecker delta Tensor … 88
- 4.5 Metric Tensor … 90
- 4.6 Definitions Involving Special Tensors … 96
  - 4.6.1 Dot Product … 97
  - 4.6.2 Magnitude of Vector … 98
  - 4.6.3 Angle between Vectors … 98
  - 4.6.4 Cross Product … 99
  - 4.6.5 Scalar Triple Product … 100
  - 4.6.6 Vector Triple Product … 101
  - 4.6.7 Determinant of Matrix … 102
  - 4.6.8 Length … 102
  - 4.6.9 Area … 103
  - 4.6.10 Volume … 105
- 4.7 Exercises and Revision … 107

5 Tensor Differentiation — 112

- 5.1 Christoffel Symbols … 112
- 5.2 Covariant Differentiation … 120
- 5.3 Absolute Differentiation … 129
- 5.4 Exercises and Revision … 133

6 Differential Operations — 137

- 6.1 Cartesian Coordinate System … 138
  - 6.1.1 Operators … 138
  - 6.1.2 Gradient … 138
  - 6.1.3 Divergence … 139
  - 6.1.4 Curl … 139
  - 6.1.5 Laplacian … 140
- 6.2 General Coordinate System … 140

		6.2.1	Operators 140

      6.2.1 Operators 140
      6.2.2 Gradient 140
      6.2.3 Divergence 142
      6.2.4 Curl 143
      6.2.5 Laplacian 144
  6.3 Orthogonal Coordinate System 146
      6.3.1 Operators 146
      6.3.2 Gradient 147
      6.3.3 Divergence 147
      6.3.4 Curl 147
      6.3.5 Laplacian 148
  6.4 Cylindrical Coordinate System 148
      6.4.1 Operators 148
      6.4.2 Gradient 149
      6.4.3 Divergence 149
      6.4.4 Curl 149
      6.4.5 Laplacian 149
  6.5 Spherical Coordinate System 150
      6.5.1 Operators 150
      6.5.2 Gradient 150
      6.5.3 Divergence 151
      6.5.4 Curl 151
      6.5.5 Laplacian 151
  6.6 Exercises and Revision 152

7 Tensors in Application   **155**
  7.1 Tensors in Mathematics 155
      7.1.1 Common Definitions in Tensor Notation 156
      7.1.2 Scalar Invariants of Tensors 157
      7.1.3 Common Identities in Vector and Tensor Notation 158
      7.1.4 Integral Theorems in Tensor Notation 160
      7.1.5 Examples of Using Tensor Techniques to Prove Identities 161
  7.2 Tensors in Geometry 167
      7.2.1 Riemann-Christoffel Curvature Tensor 167
      7.2.2 Bianchi Identities 171
      7.2.3 Ricci Curvature Tensor and Scalar 172
  7.3 Tensors in Science 173
      7.3.1 Infinitesimal Strain Tensor 173
      7.3.2 Stress Tensor 174
      7.3.3 Displacement Gradient Tensors 174
      7.3.4 Finger Strain Tensor 175
      7.3.5 Cauchy Strain Tensor 175
      7.3.6 Velocity Gradient Tensor 175

		7.3.7 Rate of Strain Tensor	176
		7.3.8 Vorticity Tensor	176
	7.4	Exercises and Revision	177

References   **180**

Index   **181**

Author Notes   **187**

Nomenclature

In the following list, we define the common symbols, notations and abbreviations which are used in the book as a quick reference for the reader.

∇	nabla differential operator
∇_i and ∇^i	covariant and contravariant differential operators
∇f	gradient of scalar f
$\nabla \cdot \mathbf{A}$	divergence of tensor $\mathbf{A}$
$\nabla \times \mathbf{A}$	curl of tensor $\mathbf{A}$
$\nabla^2, \partial_{ii}, \nabla_{ii}$	Laplacian operator
$\nabla \mathbf{v}, \partial_i v_j$	velocity gradient tensor
, (subscript)	partial derivative with respect to following index(es)
; (subscript)	covariant derivative with respect to following index(es)
hat (e.g. $\hat{A}_i, \hat{\mathbf{E}}_i$)	physical representation or normalized vector
bar (e.g. $\bar{u}^i, \bar{A}_i$)	transformed quantity
$\circ$	inner or outer product operator
$\perp$	perpendicular to
1D, 2D, 3D, nD	one-, two-, three-, n-dimensional
$\delta/\delta t$	absolute derivative operator with respect to t
∂_i and ∇_i	partial derivative operator with respect to i^{th} variable
$\partial_{;i}$	covariant derivative operator with respect to i^{th} variable
$[ij, k]$	Christoffel symbol of 1st kind
A	area
$\mathbf{B}, B_{ij}$	Finger strain tensor
$\mathbf{B}^{-1}, B_{ij}^{-1}$	Cauchy strain tensor
C	curve
C^n	of class n
$\mathbf{d}, d_i$	displacement vector
det	determinant of matrix
$d\mathbf{r}$	differential of position vector
ds	length of infinitesimal element of curve
$d\sigma$	area of infinitesimal element of surface
$d\tau$	volume of infinitesimal element of space
$\mathbf{e}_i$	i^{th} vector of orthonormal vector set (usually Cartesian basis set)
$\mathbf{e}_r, \mathbf{e}_\theta, \mathbf{e}_\phi$	basis vectors of spherical coordinate system
$\mathbf{e}_{rr}, \mathbf{e}_{r\theta}, \cdots, \mathbf{e}_{\phi\phi}$	unit dyads of spherical coordinate system
$\mathbf{e}_\rho, \mathbf{e}_\phi, \mathbf{e}_z$	basis vectors of cylindrical coordinate system
$\mathbf{e}_{\rho\rho}, \mathbf{e}_{\rho\phi}, \cdots, \mathbf{e}_{zz}$	unit dyads of cylindrical coordinate system
$\mathbf{E}, E_{ij}$	first displacement gradient tensor
$\mathbf{E}_i, \mathbf{E}^i$	i^{th} covariant and contravariant basis vectors
$\underline{\mathbf{E}}_i$	i^{th} orthonormalized covariant basis vector

Eq./Eqs.	Equation/Equations
g	determinant of covariant metric tensor
$\mathbf{g}$	metric tensor
g_{ij}, g^{ij}, g_i^j	covariant, contravariant and mixed metric tensor or its components
g_{11}, g_{12}, g_{22}	coefficients of covariant metric tensor
g^{11}, g^{12}, g^{22}	coefficients of contravariant metric tensor
h_i	scale factor for i^{th} coordinate
iff	if and only if
J	Jacobian of transformation between two coordinate systems
$\mathbf{J}$	Jacobian matrix of transformation between two coordinate systems
$\mathbf{J}^{-1}$	inverse Jacobian matrix of transformation
L	length of curve
$\mathbf{n}, n_i$	normal vector to surface
P	point
$P(n,k)$	k-permutations of n objects
q^i	i^{th} coordinate of orthogonal coordinate system
$\mathbf{q}_i$	i^{th} unit basis vector of orthogonal coordinate system
$\mathbf{r}$	position vector
$\mathcal{R}$	Ricci curvature scalar
R_{ij}, R^i_j	Ricci curvature tensor of 1st and 2nd kind
R_{ijkl}, R^i_{jkl}	Riemann-Christoffel curvature tensor of 1st and 2nd kind
r, θ, ϕ	coordinates of spherical coordinate system
S	surface
$\mathbf{S}, S_{ij}$	rate of strain tensor
$\bar{\mathbf{S}}, \bar{S}_{ij}$	vorticity tensor
t	time
T (superscript)	transposition of matrix
$\mathbf{T}, T_i$	traction vector
tr	trace of matrix
u^i	i^{th} coordinate of general coordinate system
$\mathbf{v}, v_i$	velocity vector
V	volume
w	weight of relative tensor
x_i, x^i	i^{th} Cartesian coordinate
x'_i, x_i	i^{th} Cartesian coordinate of particle at past and present times
x, y, z	coordinates of 3D space (mainly Cartesian)
$\boldsymbol{\gamma}, \gamma_{ij}$	infinitesimal strain tensor
$\dot{\boldsymbol{\gamma}}$	rate of strain tensor
Γ^k_{ij}	Christoffel symbol of 2nd kind
$\boldsymbol{\delta}$	Kronecker delta tensor
$\delta_{ij}, \delta^{ij}, \delta_i^j$	covariant, contravariant and mixed ordinary Kronecker delta
$\delta^{ij}_{kl}, \delta^{ijk}_{lmn}, \delta^{i_1...i_n}_{j_1...j_n}$	generalized Kronecker delta in 2D, 3D and nD space
$\boldsymbol{\Delta}, \Delta_{ij}$	second displacement gradient tensor

ϵ_{ij}, ϵ_{ijk}, $\epsilon_{i_1...i_n}$	covariant relative permutation tensor in 2D, 3D and nD space
ϵ^{ij}, ϵ^{ijk}, $\epsilon^{i_1...i_n}$	contravariant relative permutation tensor in 2D, 3D and nD space
$\underline{\epsilon}_{ij}$, $\underline{\epsilon}_{ijk}$, $\underline{\epsilon}_{i_1...i_n}$	covariant absolute permutation tensor in 2D, 3D and nD space
$\underline{\epsilon}^{ij}$, $\underline{\epsilon}^{ijk}$, $\underline{\epsilon}^{i_1...i_n}$	contravariant absolute permutation tensor in 2D, 3D and nD space
ρ, ϕ	coordinates of plane polar coordinate system
ρ, ϕ, z	coordinates of cylindrical coordinate system
$\boldsymbol{\sigma}$, σ_{ij}	stress tensor
$\boldsymbol{\omega}$	vorticity tensor
Ω	region of space

Chapter 1
Preliminaries

In this introductory chapter, we provide preliminary materials about conventions and notations as well as basic facts about tensors which will be needed in the subsequent parts of the book. The chapter is therefore divided into two sections: the first is about general conventions and notations used in the book, and the second is on general background about tensors.

1.1 General Conventions and Notations

In this section, we provide general notes about the main conventions and notations used in the present book. We usually use the term "tensor" to mean tensors of all ranks including scalars (rank-0) and vectors (rank-1). However, we may also use this term as opposite to scalar and vector, i.e. tensor of rank-n where $n > 1$. In almost all cases, the meaning should be obvious from the context. We note that in the present book all tensors of all ranks and types are assumed to be real quantities, i.e. they have real rather than imaginary or complex components.

We use non-indexed lower case light face italic Latin letters (e.g. f and h) to label scalars, while we use non-indexed lower or upper case bold face non-italic Latin letters (e.g. **a** and **A**) to label vectors in symbolic notation. The exception to this is the basis vectors where indexed bold face lower or upper case non-italic symbols (e.g. $\mathbf{e}_1$ and $\mathbf{E}^i$) are used. However, there should be no confusion or ambiguity about the meaning of any one of these symbols. We also use non-indexed upper case bold face non-italic Latin letters (e.g. **A** and **B**) to label tensors of rank > 1 in symbolic notation. Since matrices in this book are supposed to represent rank-2 tensors, they also follow the rules of labeling tensors symbolically by using non-indexed upper case bold face non-italic Latin letters. We note that in a few cases in the final chapter (see § 7.3) we used boldface and indexed light face Greek symbols to represent particular tensors, which are commonly labeled in the literature by these symbols, to keep with the tradition.

Indexed light face italic Latin symbols (e.g. a_i and B_i^{jk}) are used in this book to denote tensors of rank > 0 in their explicit tensor form, i.e. index notation. Such symbols may also be used to denote the components of these tensors. The meaning is usually transparent and can be identified from the context if it is not declared explicitly. Tensor indices in this book are lower case Latin letters which may be taken preferably from the middle of the Latin alphabet (such as i, j and k) for the free indices and from the beginning of the Latin alphabet (such as a and b) for the dummy indices. We also use numbered indices, such as $(i_1, i_2, \ldots, i_k)$, for this purpose when the number of tensor indices is variable. Numbers are also used as indices in some occasions (e.g. ϵ_{12}) for obvious purposes such as making statements about particular components.

Partial derivative symbol with a subscript index (e.g. ∂_i) is used to denote partial differentiation with respect to the ith variable, that is:

$$\partial_i = \frac{\partial}{\partial x^i} \tag{1}$$

However, we should note that in this book we generalize partial derivative notation so that ∂_i symbolizes partial derivative with respect to the u^i coordinate of general coordinate systems and not just Cartesian coordinates which are usually denoted by x_i or x^i. The type of coordinates, being Cartesian or general or otherwise, will be determined by the context which should be obvious in all cases.

Similarly, we use partial derivative symbol with a twice-repeated index to denote the Laplacian operator, that is:[1]

$$\partial_{ii} = \partial_i \partial_i = \nabla^2 \tag{2}$$

Partial derivative symbol with a coordinate label subscript, rather than an index, is also used to denote partial differentiation with respect to that spatial variable. For instance:

$$\partial_r = \frac{\partial}{\partial r} \tag{3}$$

is used to denote the partial derivative with respect to the radial coordinate r in spherical coordinate systems which are identified by the spatial variables (r, θ, ϕ). It should be obvious that in notations like ∂_r the subscript is used as a label rather than an index and hence it does not follow the rules of tensor indices which will be discussed later (see § 1.2).

Following the widely used convention, a subscript comma preceding a subscript index (e.g. $A_{k,i}$) is used to denote partial differentiation with respect to the spatial coordinate which is indexed by the symbol that follows the comma. For example, $f_{,i}$ and $A^{jk}_{,i}$ are used to represent the partial derivative of the scalar f and rank-2 tensor A^{jk} with respect to the ith coordinate, that is:

$$f_{,i} = \partial_i f \qquad A^{jk}_{,i} = \partial_i A^{jk} \tag{4}$$

We also follow the common convention of using a subscript semicolon preceding a subscript index (e.g. $A_{kl;i}$) to symbolize the operation of covariant differentiation with respect to the ith coordinate (see § 5.2). The semicolon notation may also be attached to the normal differential operators for the same purpose. For example, $\nabla_{;i}$ and $\partial_{;i}$ symbolize covariant differential operators with respect to the ith variable.

In this regard, we should remark that more than one index may follow the comma and semicolon in these notations to represent multiple partial and covariant differentiation with respect to the indexed variables according to the stated order of the indices. For example, $A^i_{,jk}$ is used to represent the mixed second order partial derivative of the tensor A^i with respect to the jth and k^{th} coordinates, while $B_{ji;km}$ is used to represent the mixed second

---

[1] This is the Cartesian form. For the other forms, the reader is referred to § 6.

1.1 General Conventions and Notations

order covariant derivative of the tensor B_{ji} with respect to the kth and m^{th} coordinates, that is:

$$A^i{}_{,jk} = \partial_k \left(\partial_j A^i \right) \qquad\qquad B_{ji;km} = \nabla_{;m} \left(\nabla_{;k} B_{ji} \right) \qquad (5)$$

We also note that in a few occasions superscripts, rather than subscripts, comma and semi-colon preceding superscript index (e.g. $f^{,i}$ and $A_i{}^{;j}$) are used to represent contravariant partial derivative and contravariant tensor derivative respectively. Superscripted differential operators (e.g. ∂^i and $\nabla^{;i}$) are also used occasionally to represent these differential operators in their contravariant form. A matter related to tensor differentiation is that we follow the conventional notation $\frac{\delta}{\delta t}$ to represent the intrinsic derivative, which is also known as the absolute derivative, with respect to the variable t along a given curve, as will be discussed in § 5.3.

Due to the restriction that we impose of using real (as opposite to imaginary and complex) quantities exclusively in this book, all arguments of real-valued functions which are not defined for negative quantities, like square roots and logarithmic functions, are assumed to be non-negative by taking the absolute value, if necessary, without using the absolute value symbol. This is to simplify the notation and avoid potential confusion with the determinant notation. So, $\sqrt{g}$ means $\sqrt{|g|}$ and $\ln(g)$ means $\ln(|g|)$.

We follow the summation convention which is widely used in the literature of tensor calculus and its applications. However, the summation symbol (i.e. Σ) is used in a few cases where a summation operation is needed but the conditions of the summation convention do not apply or there is an ambiguity about them, e.g. when an index is repeated more than twice or when a summation index is not repeated visually because it is part of a squared symbol. In a few other cases, where a twice-repeated index that complies with the conditions of the summation convention does not imply summation and hence the summation convention do not apply, we clarified the situation by adding comments like "no sum on index". We may also add a "no sum" comment in some cases where the conditions of the summation convention do not apply technically but the expression may be misleading since it contains a repetitive index, e.g. when both indices are of the same variance type in a general coordinate system such as g^{ii} or when one of the apparent indices is in fact a label for a scalar rather than a variable index such as $|\mathbf{E}_i|$ or h_j.

All the transformation equations in the present book are continuous and real, and all the derivatives are continuous over their intended domain. Based on the well known continuity condition of differential calculus, the individual differential operators in the second (and higher) order partial derivatives with respect to different indices are commutative, that is:

$$\partial_i \partial_j = \partial_j \partial_i \qquad (6)$$

We generally assume that this continuity condition is satisfied and hence the order of the partial differential operators in these mixed second order partial derivatives does not matter.

We use vertical bars (i.e. |::|) to symbolize determinants and square brackets (i.e. [::]) to symbolize matrices. This applies when these symbols contain arrays of objects; otherwise they have their normal meaning according to the context, e.g. bars embracing a vector such as $|\mathbf{v}|$ mean modulus of the vector. Also, we use indexed square brackets (such as $[\mathbf{A}]^i$ and $[\nabla f]_i$) to denote the ith component of vectors in their symbolic or vector notation. For tensors of higher rank, more than one index are used to denote their components, e.g. $[\mathbf{A}]_{ij}$ represents the ij^{th} component of the rank-2 tensor $\mathbf{A}$.

We finally should remark that although we generally talk about nD spaces, our main focus is the low dimensionality spaces (mostly 2D and 3D) especially with regard to coordinate systems and hence some of the statements may apply only to these low dimensionality spaces although the statements are given in the context of nD spaces. In most cases, such statements can be generalized simply by adding extra conditions or by a slight modification to the phrasing and terminology.

1.2 General Background about Tensors

A tensor is an array of mathematical objects (usually numbers or functions) which transforms according to certain rules under coordinates change. In an nD space, a tensor of rank-k has n^k components which may be specified with reference to a given coordinate system. Accordingly, a scalar, such as temperature, is a rank-0 tensor with (assuming a 3D space) $3^0 = 1$ component, a vector, such as force, is a rank-1 tensor with $3^1 = 3$ components, and stress is a rank-2 tensor with $3^2 = 9$ components. In Figure 1 we graphically illustrate the structure of a rank-3 tensor in a 3D space.

The n^k components of a rank-k tensor in an nD space are identified by k distinct integer indices (e.g. i, j, k) which are attached, according to the commonly-employed tensor notation, as superscripts or subscripts or a mix of these to the right side of the symbol utilized to label the tensor, e.g. A_{ijk}, A^{ijk} and A_i^{jk}. Each tensor index takes all the values over a predefined range of dimensions such as 1 to n in the above example of an nD space. In general, all tensor indices have the same range, i.e. they are uniformly dimensioned.[2] When the range of tensor indices is not stated explicitly, it is usually assumed to range over the values $1, 2, 3$. However, the range must be stated explicitly or implicitly to avoid ambiguity.

The characteristic property of tensors is that they satisfy the principle of invariance under certain coordinate transformations. Therefore, formulating the fundamental laws of physics in a tensor form ensures that they are form-invariant, and hence they are objectively representing the physical reality and do not depend on the observer and his coordinate system. Having the same form in different coordinate systems may also be labeled as being covariant although this term is usually used for a different meaning in tensor calculus, as will be explained in § 3.1.1.

While tensors of rank-0 are generally represented in a common form of light face non-indexed italic symbols like f and h, tensors of rank ≥ 1 are represented in several forms and notations, the main ones are the index-free notation, which may also be called the direct or symbolic or Gibbs notation, and the indicial notation which is also called the

---

[2] This applies to the common cases of tensor applications, but there are instances (e.g. in differential geometry of curves and surfaces) of tensors which are not uniformly dimensioned because the tensor is related to two spaces with different dimensions such as a 2D surface embedded in a 3D space.

1.2 General Background about Tensors

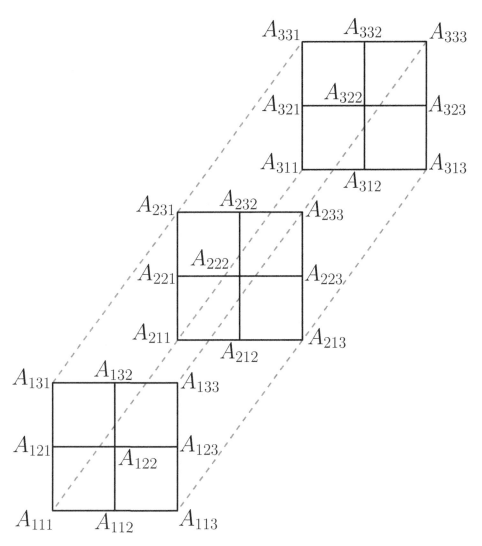

Figure 1: Graphical illustration of a rank-3 tensor A_{ijk} in a 3D space, i.e. each one of i, j, k ranges over $1, 2, 3$.

index or component or tensor notation. The first is a geometrically oriented notation with no reference to a particular coordinate system and hence it is intrinsically invariant to the choice of coordinate systems, whereas the second takes an algebraic form based on components identified by indices and hence the notation is suggestive of an underlying coordinate system, although being a tensor makes it form-invariant under certain coordinate transformations and therefore it possesses certain invariant properties. The index-free notation is usually identified by using bold face non-italic symbols, like **a** and **B**, while the indicial notation is identified by using light face indexed italic symbols such as a^i and B_{ij}. It is noteworthy that although rank-0 and rank-1 tensors are, respectively, scalars and vectors, not all scalars and vectors (in their generic sense) are tensors of these ranks. Similarly, rank-2 tensors are normally represented by square matrices but not all square matrices represent rank-2 tensors.

1.2 General Background about Tensors

Tensors can be combined through common algebraic operations such as addition and multiplication. Tensor term is a product of tensors including scalars and vectors and may consist of a single tensor which can be regarded as a multiple of unity. Tensor expression is an algebraic sum of tensor terms which may be a trivial sum in the case of a single term. Tensor equality is an equality of two tensor terms and/or expressions. An index that occurs once in a tensor term is a free index while an index that occurs twice in a tensor term is a dummy or bound index.

The order of a tensor is identified by the number of its indices. For example, A^i_{jk} is a tensor of order 3 while B_{km} is a tensor of order 2. The order of the tensor normally identifies its rank as well and hence A^i_{jk} is of rank-3 and B_{km} is of rank-2. However, when the operation of contraction of indices (see § 3.2.4) takes place once or more, the order of the tensor is not affected but its rank is reduced by two for each contraction operation. Hence, the order of a tensor is equal to the number of all of its indices including the dummy indices, while the rank is equal to the number of its free indices only. Accordingly, A^{ab}_{abmn} is of order 6 and rank-2 while B^a_{ai} is of order 3 and rank-1. We note that many authors follow different conventions such as using "order" as equivalent to what we call "rank".

Tensors whose all indices are subscripts, like A_{ij}, are called covariant, while tensors whose all indices are superscripts, like A^k, are called contravariant. Tensors with both types of indices, like A^{amn}_{ak}, are called mixed type. Subscript indices, rather than subscripted tensors, are also described as covariant and superscript indices are described as contravariant. The zero tensor is a tensor whose all components are zero. The unit tensor or unity tensor, which is usually defined for rank-2 tensors, is a tensor whose all elements are zero except those with identical values of all indices (e.g. A_{11} or B^{33}) which are assigned the value 1.

There are general rules that govern the manipulation of tensors and hence they should be observed in the handling of mathematical expressions and calculations of tensor calculus. One of these rules is that no tensor index is allowed to occur more than twice in a legitimate tensor term. However, we follow in this assertion the common literature of tensor calculus which represents the ordinary use of repeated indices in tensor terms. In fact, there are many instances in the literature of tensor calculus where indices are legitimately repeated more than twice in a single term. The bottom line is that as long as the tensor expression makes sense and the intention is clear, such repetitions should be allowed with no need to take special precautions like using parentheses as done by some authors. In particular, the forthcoming summation convention will not apply automatically in such cases although summation on such indices, if needed, can be carried out explicitly, by using the summation symbol Σ, or by a special declaration of such intention similar to the summation convention which is usually restricted to the twice-repeated indices.

Regarding the aforementioned summation convention, according to this convention which is widely used in the literature of tensor calculus including the present book, dummy indices imply summation over their range. More clearly, a twice-repeated variable (i.e. not numeric) index in a single term implies a sum of terms equal in number to the range of the repeated index. Hence, in a 3D space we have:

$$A^{aj}_a = A^{1j}_1 + A^{2j}_2 + A^{3j}_3 \tag{7}$$

1.2 General Background about Tensors

while in a 4D space we have:

$$B_i + C_{ia}D^a = B_i + C_{i1}D^1 + C_{i2}D^2 + C_{i3}D^3 + C_{i4}D^4 \tag{8}$$

We note that although the twice-repeated index should be in the same term for the summation convention to apply, it does not matter if the two indices occur in one tensor or in two tensors, as seen in the last example where the summation convention applies to a.

We should also remark that there are many cases in the mathematics of tensors where a repeated index is needed but with no intention of summation, such as using A_{ii} or A_j^j to mean the components of the tensor $\mathbf{A}$ whose indices have identical value like A_{11} and A_2^2. So to avoid confusion, when the dummy indices in a particular case do not imply summation, the situation must be clarified by enclosing such indices in parentheses or by underscoring or by using upper case letters with declaration of these conventions, or by adding a clarifying comment like "no summation over repeated indices". These precautions are obviously needed if the summation convention is adopted in general but it does not apply in some exceptional cases where repeated indices are needed in the notation with no implication of summation.

Another rule of tensors is that a free index should be understood to vary over its range (e.g. $1, \ldots, n$) which is determined by the space dimension and hence it should be interpreted as saying "for all components represented by the index". Therefore, a free index represents a number of terms or expressions or equalities equal to the number of the allowed values of its range. For example, when i and j can vary over the range $1, \ldots, n$ then the expression $A_i + B_i$ represents n separate expressions while the equation $A_i^j = B_i^j$ represents $n \times n$ separate equations which represent the combination of all possible n values of i with all possible n values of j, that is:

$$A_1 + B_1, \quad A_2 + B_2, \quad \cdots, \quad A_n + B_n \tag{9}$$

$$A_1^1 = B_1^1, \quad A_1^2 = B_1^2, \quad A_2^1 = B_2^1, \quad \cdots, \quad A_n^n = B_n^n \tag{10}$$

Also, each tensor index should conform to one of the forthcoming variance transformation rules as given by Eqs. 71 and 72, i.e. it is either covariant or contravariant. For orthonormal Cartesian coordinate systems, the two variance types (i.e. covariant and contravariant) do not differ because the metric tensor is given by the Kronecker delta (refer to § 4.1 and 4.5) and hence any index can be upper or lower although it is common to use lower indices in this case. We note that orthonormal vectors mean a set of vectors which are mutually orthogonal and each one is of unit length, while orthonormal coordinate system means a coordinate system whose basis vector set is orthonormal at all points of the space where the system is defined (see § 2.6). The orthonormality of vectors may be expressed mathematically by:

$$\mathbf{V}_i \cdot \mathbf{V}_j = \delta_{ij} \qquad \text{or} \qquad \mathbf{V}^i \cdot \mathbf{V}^j = \delta^{ij} \tag{11}$$

where the indexed δ is the Kronecker delta symbol and the indexed $\mathbf{V}$ symbolizes a vector in the set.

In this context, we should remark that for tensor invariance, a pair of dummy indices involved in summation should in general be complementary in their variance type, i.e. one covariant and the other contravariant. However, for orthonormal Cartesian systems the two variance types are the same and hence when both dummy indices are covariant or both are contravariant it should be understood as an indication that the underlying coordinate system is orthonormal Cartesian if the possibility of an error is excluded.

As indicated earlier, tensor order is equal to the number of its indices while tensor rank is equal to the number of its free indices. Hence, scalars (terms, expressions and equalities) have no free index since they are of rank-0, and vectors have a single free index while rank-2 tensors have exactly two free indices. Similarly, rank-n tensors have exactly n free indices. The dimension of a tensor is determined by the range taken by its indices which represents the number of dimensions of the underlying space. For example, in a 3D space the tensor A_i^j is of rank-2 because it possesses exactly two free indices but it is of dimension three since each one of its free indices range over the values $1, 2, 3$ and hence it may be represented by a 3×3 matrix. However, in a 4D space this rank-2 tensor will be represented by a 4×4 matrix since its two free indices range over $1, 2, 3, 4$.

The rank of all terms in legitimate tensor expressions and equalities must be the same and hence:

$$A_i^j - B_i^j \qquad \text{and} \qquad A_i^j = C_i^j \tag{12}$$

are legitimate but:

$$A_i^j - B_i \qquad \text{and} \qquad A_i^j = C^j \tag{13}$$

are illegitimate. Moreover, each term in valid tensor expressions and equalities must have the same set of free indices (e.g. i, j, k) and hence:

$$A_i^{jk} - B_i^{jk} \qquad \text{and} \qquad A_{im}^j = C_{im}^j \tag{14}$$

are legitimate but:

$$A_i^{jk} - B_i^{jm} \qquad \text{and} \qquad A_{im}^j = C_{in}^j \tag{15}$$

are illegitimate although they are all of the same rank.

Also, a free index should keep its variance type in every term in valid tensor expressions and equations, i.e. it must be covariant in all terms or contravariant in all terms, and hence:

$$A_{iq}^{mk} - B_{iq}^{mk} \qquad \text{and} \qquad A_i^j = C_i^j \tag{16}$$

are legitimate but:

$$A_i^j - B_j^i \qquad \text{and} \qquad A_{in}^j = C_i^{jn} \tag{17}$$

are illegitimate although they are all of the same rank and have the same set of free indices. We also note that the order of the tensor indices in legitimate tensor expressions and equalities should be the same and hence:

$$A_{km} + C_{km} \qquad \text{and} \qquad D^j{}_{in} = E^j{}_{in} \tag{18}$$

1.2 General Background about Tensors

are legitimate but:

$$A^i_{jk} + C^i_{kj} \qquad \text{and} \qquad D^{nm}_i = E^{mn}_i \qquad (19)$$

are illegitimate although they are all of the same rank, have the same set of free indices and have the same variance type.

We remark that expressions and equalities like:

$$A_{ij} + B_{ji} \qquad \text{and} \qquad A^{iq}_k = B^{qi}_k - C^{iq}_k \qquad (20)$$

are common in the literature of tensor calculus and its applications which may suggest that the order of indices (or another one of the aforementioned features of the indicial structure) in tensor expressions and equalities is not important. However, expressions and equalities like these refer to the individual components of these tensors and hence they are of scalar, rather than tensor, nature. Hence the expression $A_{ij} + B_{ji}$ means adding the value of the component A_{ij} of tensor **A** to the value of the component B_{ji} of tensor **B** and not adding the tensor **A** to the tensor **B**. Similarly, the equality $A^{iq}_k = B^{qi}_k - C^{iq}_k$ means subtracting the value of the component C^{iq}_k of tensor **C** from the value of the component B^{qi}_k of tensor **B** to obtain the value of the component A^{iq}_k of tensor **A**. We may similarly write things like $A_i = B^i$ or $A_i = B_j$ to mean the equality of the values of these components and not tensor equality.

As we will see (refer to § 3.1.1), the indicial notation of tensors is made with reference to a set of basis vectors. For example, when we write A^i_{jk} as a tensor we mean $A^i_{jk}\mathbf{E}_i\mathbf{E}^j\mathbf{E}^k$. This justifies all the above rules about the indicial structure (rank, set of free indices, variance type and order of indices) of tensor terms involved in tensor expressions and equalities because this structure is based on a set of basis vectors in a certain order. Therefore, an expression like $A^i_{jk} + B^i_{jk}$ means $A^i_{jk}\mathbf{E}_i\mathbf{E}^j\mathbf{E}^k + B^i_{jk}\mathbf{E}_i\mathbf{E}^j\mathbf{E}^k$ and an equation like $A^i_{jk} = B^i_{jk}$ means $A^i_{jk}\mathbf{E}_i\mathbf{E}^j\mathbf{E}^k = B^i_{jk}\mathbf{E}_i\mathbf{E}^j\mathbf{E}^k$. However, as seen above it is common in the literature of tensor calculus that the tensor notation like A^i_{jk} is used to label the components and hence the above rules are not respected (e.g. $B_{ij} + C_{ji}$ or $\epsilon_{ij} = \epsilon^{ij}$) because these components are scalars in nature and hence these expressions and equalities do not refer to the vector basis. In this context, we should remark that an additional condition may be imposed on the indicial structure that is all the indices of a tensor should refer to a single vector basis set and hence a tensor cannot be indexed in reference to two different basis sets simultaneously.[3]

While free indices should be named uniformly in all terms of tensor expressions and equalities, dummy indices can be named in each term independently and hence:

$$A^a_{ak} + B^b_{bk} + C^c_{ck} \qquad \text{and} \qquad D^j_i = E^{ja}_{ia} + F^{jb}_{ib} \qquad (21)$$

---

[3] A tensor of different dimensionality (as in differential geometry) may not be compliant with this condition and hence we may add an extra condition that all the indices should refer to a single vector basis set *for any particular space*. Alternatively, a tensor like this can be seen as a tensor in reference to each individual type of indices separately but it is not a tensor simultaneously to both types of indices where the *simultaneously* condition takes care of this.

1.2 General Background about Tensors

are legitimate. The reason is that a dummy index represents a sum in its own term with no reach or presence into other terms. Despite the above restriction on the free indices, a free index in an expression or equality can be renamed uniformly and thoroughly using a different symbol, as long as this symbol is not already in use, assuming that both symbols vary over the same range, i.e. they have the same dimension. For example, we can change:

$$A_i^{jk} + B_i^{jk} \qquad \text{to} \qquad A_p^{jk} + B_p^{jk} \qquad (22)$$

and change:

$$A_{im}^j = C_{im}^j - D_{im}^j \qquad \text{to} \qquad A_{pm}^j = C_{pm}^j - D_{pm}^j \qquad (23)$$

as long as i and p have the same range and p is not already in use as an index for another purpose in that context. As indicated, the change should be thorough and hence all occurrences of the index i in that context, which may include other expressions and equalities, should be subject to that change. Regarding the dummy indices, they can be replaced by another symbol which is not present (as a free or dummy index) in their term as long as there is no confusion with a similar symbol in that context.

Indexing is generally distributive over the terms of tensor expressions and equalities. For example, we have:

$$[\mathbf{A} + \mathbf{B}]_i = [\mathbf{A}]_i + [\mathbf{B}]_i \qquad (24)$$

and

$$[\mathbf{A} = \mathbf{B}]_i \quad \Longleftrightarrow \quad [\mathbf{A}]_i = [\mathbf{B}]_i \qquad (25)$$

Unlike scalars and tensor components, which are essentially scalars in a generic sense, operators cannot in general be freely reordered in tensor terms. Therefore, we have the following legitimate equalities:

$$fh = hf \qquad \text{and} \qquad A_i B^j = B^j A_i \qquad (26)$$

but we cannot equate $\partial_i A_j$ to $A_j \partial_i$ since in general we have:

$$\partial_i A_j \neq A_j \partial_i \qquad (27)$$

This should be obvious because $\partial_i A_j$ means that ∂_i is operating on A_j but $A_j \partial_i$ means that ∂_i is operating on something else and A_j just multiplies the result of this operation.

As seen above, the order of the indices[4] of a given tensor is important and hence it should be observed and clarified, because two tensors with the same set of free indices and with the same indicial structure that satisfies the aforementioned rules but with different indicial order are not equal in general. For example, A_{ijk} is not equal to A_{jik} unless $\mathbf{A}$ is symmetric with respect to the indices i and j (refer to § 3.1.5). Similarly, B^{mln} is not equal to B^{lmn} unless $\mathbf{B}$ is symmetric in its indices l and m. The confusion about the order of indices occurs specifically in the case of mixed type tensors such as A^i_{jk} which may not

---
[4] The "order" here means "arrangement" and hence it should not be confused with the order of tensor as defined above in the same context as tensor rank.

be clear since the order can be *ijk* or *jik* or *jki*. Spaces are usually used in this case to clarify the order. For example, the latter tensor is symbolized as $A^i{}_{jk}$ if the order of the indices is *ijk*, and as $A_j{}^i{}_k$ if the order of the indices is *jik* while it is symbolized as $A_{jk}{}^i$ if the order of the indices is *jki*. Dots may also be used in such cases to indicate, more explicitly, the order of the indices and remove any ambiguity. For example, if the indices i, j, k of the tensor **A**, which is covariant in i and k and contravariant in j, are of that order, then **A** may be symbolized as $A_i{}^j{}_{.k}$ where the dot between i and k indicates that j is in the middle.

We note that in many places in this book (like many other books of tensor calculus) and mostly for the sake of convenience in typesetting, the order of the indices of mixed type tensors is not clarified by spacing or by inserting dots. This commonly occurs when the order of the indices is irrelevant in the given context (e.g. any order satisfies the intended purpose) or when the order is clear. Sometimes, the order of the indices may be indicated implicitly by the alphabetical order of the selected indices, e.g. writing A_i^{jk} to mean $A_i^{.jk}$.

Finally, scalars, vectors and tensors may be defined on a single point of the space or on a set of separate points. They may also be defined over an extended continuous region (or regions) of the space. In the latter case we have scalar fields, vector fields and tensor fields, e.g. temperature field, velocity field and stress field respectively. Hence, a "field" is a function of coordinates which is defined over a given region of the space. As stated earlier, "tensor" may be used in a general sense to include scalar and vector and hence "tensor field" may include all the three types.

1.3 Exercises and Revision

1.1 Differentiate between the symbols used to label scalars, vectors and tensors of rank > 1.
1.2 What the comma and semicolon in $A^{jk}_{,i}$ and $A_{k;i}$ mean?
1.3 State the summation convention and explain its conditions. To what type of indices this convention applies?
1.4 What is the number of components of a rank-3 tensor in a 4D space?
1.5 A symbol like B_i^{jk} may be used to represent tensor or its components. What is the difference between these two representations? Do the rules of indices apply to both representations or not? Justify your answer.
1.6 What is the meaning of the following symbols: ∇, ∂_j, ∂_{kk}, ∇^2, ∂_ϕ, $h_{,jk}$, $A^i_{;n}$, ∂^n, $\nabla^{;k}$ and $C_{i;km}$?
1.7 What is the difference between symbolic notation and indicial notation? For what type of tensors these notations are used? What are the other names given to these types of notation?
1.8 "The characteristic property of tensors is that they satisfy the principle of invariance under certain coordinate transformations". Does this mean that the components of tensors are constant? Why this principle is very important in physical sciences?
1.9 State and explain all the notations used to represent tensors of all ranks (rank-0, rank-1, rank-2, etc.). What are the advantages and disadvantages of using each one

1.3 Exercises and Revision

of these notations?

1.10 State the continuity condition that should be met if the equality: $\partial_i \partial_j = \partial_j \partial_i$ is to be correct.

1.11 Explain the difference between free and bound tensor indices. Also, state the rules that govern each one of these types of index in tensor terms, expressions and equalities.

1.12 Explain the difference between the order and the rank of tensors and link this to the free and dummy indices.

1.13 What is the difference between covariant, contravariant and mixed type tensors? Give an example for each.

1.14 What is the meaning of "unit" and "zero" tensors? What is the characteristic feature of these tensors with regard to the value of their components?

1.15 What is the meaning of "orthonormal vector set" and "orthonormal coordinate system"? State any relevant mathematical condition.

1.16 What is the rule that governs the pair of dummy indices involved in summation regarding their variance type in general coordinate systems? Which type of coordinate system is exempt of this rule and why?

1.17 State all the rules that govern the indicial structure of tensors involved in tensor expressions and equalities (rank, set of free indices, variance type and order of indices).

1.18 How many equalities that the following equation contains assuming a 4D space: $B_i^k = C_i^k$? Write all these equalities explicitly, i.e. $B_1^1 = C_1^1$, $B_1^2 = C_1^2$, etc.

1.19 Which of the following tensor expressions is legitimate and which is not, giving detailed explanation in each case?

$$A_i^k - B_i, \qquad C_a^a + D_m^n - B_b^b, \qquad a + B, \qquad S_{cdk}^{cdj} + F_{abk}^{abj}$$

1.20 Which of the following tensor equalities is legitimate and which is not, giving detailed explanation in each case?

$$A_i^{\cdot n} = B_{\cdot i}^n, \qquad D = S_c^c + N_{ba}^{ab}, \qquad 3a + 2b = J_a^a, \qquad B_k^m = C_m^k, \qquad B_j = 3c - D_j$$

1.21 Explain why the indicial structure (rank, set of free indices, variance type and order of indices) of tensors involved in tensor expressions and equalities are important referring in your explanation to the vector basis set to which the tensors are referred. Also explain why these rules are not observed in the expressions and equalities of tensor components.

1.22 Why free indices should be named uniformly in all terms of tensor expressions and equalities while dummy indices can be named in each term independently?

1.23 What are the rules that should be observed when replacing the symbol of a free index with another symbol? What about replacing the symbols of dummy indices?

1.24 Why in general we have: $\partial_i A_j \neq A_j \partial_i$? What are the situations under which the following equality is valid: $\partial_i A_j = A_j \partial_i$?

1.25 What is the difference between the order of a tensor and the order of its indices?

1.26 In which case A_{ijk} is equal to A_{ikj}? What about A_{ijk} and A^{ikj}?

1.27 What are the rank, order and dimension of the tensor A_{jk}^i in a 3D space? What about the scalar f and the tensor A_{abjn}^{abm} from the same perspectives?

1.3 Exercises and Revision

1.28 What is the order of indices in $A_j{}^i{}_k$? Insert a dot in this symbol to make the order more explicit.

1.29 Why the order of indices of mixed tensors may not be clarified by using spaces or inserting dots?

1.30 What is the meaning of "tensor field"? Is A^i a tensor field considering the spatial dependency of A^i and the meaning of "tensor"?

Chapter 2
Spaces, Coordinate Systems and Transformations

The focus of this chapter is coordinate systems, their types and transformations as well as some general properties of spaces which are needed for the development of the concepts and techniques of tensor calculus in the present and forthcoming chapters. The chapter also includes other sections which are intimately linked to these topics.

2.1 Spaces

A Riemannian space is a manifold characterized by the existing of a symmetric rank-2 tensor called the metric tensor. The components of this tensor, which can be in covariant form g_{ij} or contravariant form g^{ij}, as well as mixed form g^i_j, are continuous variable functions of coordinates in general, that is:

$$g_{ij} = g_{ij}(u^1, u^2, \ldots, u^n) \qquad (28)$$
$$g^{ij} = g^{ij}(u^1, u^2, \ldots, u^n) \qquad (29)$$
$$g^i_j = g^i_j(u^1, u^2, \ldots, u^n) \qquad (30)$$

where the indexed u symbolizes general coordinates. This tensor facilitates, among other things, the generalization of the concept of length in general coordinate systems where the length of an infinitesimal element of arc, ds, is defined by:

$$(ds)^2 = g_{ij} du^i du^j \qquad (31)$$

In the special case of a Euclidean space coordinated by an orthonormal Cartesian system, the metric becomes the identity tensor, that is:

$$g_{ij} = \delta_{ij} \qquad g^{ij} = \delta^{ij} \qquad g^i_j = \delta^i_j \qquad (32)$$

More details about the metric tensor and its significance and roles will be given in § 4.5.

The metric of a Riemannian space may be called the Riemannian metric. Similarly, the geometry of the space may be described as the Riemannian geometry. All spaces dealt with in the present book are Riemannian with well-defined metrics. As we will see, an nD manifold is Euclidean *iff* the Riemann-Christoffel curvature tensor vanishes identically (see § 7.2.1); otherwise the manifold is curved to which the general Riemannian geometry applies. In metric spaces, the physical quantities are independent of the form of description, being covariant or contravariant, as the metric tensor facilitates the transformation between the different forms; hence making the description objective.

2.1 Spaces

A manifold, such as a 2D surface or a 3D space, is called "flat" if it is possible to find a coordinate system for the manifold with a diagonal metric tensor whose all diagonal elements are ±1; the space is called "curved" otherwise. More formally, an nD space is described as flat space *iff* it is possible to find a coordinate system for which the length of an infinitesimal element of arc ds is given by:

$$(ds)^2 = \zeta_1(du^1)^2 + \zeta_2(du^2)^2 + \ldots + \zeta_n(du^n)^2 = \sum_{i=1}^{n} \zeta_i(du^i)^2 \qquad (33)$$

where the indexed ζ are ±1 while the indexed u are the coordinates of the space. For the space to be flat (i.e. globally not just locally), the condition given by Eq. 33 should apply all over the space and not just at certain points or regions.

An example of flat space is the 3D Euclidean space which can be coordinated by an orthonormal Cartesian system whose metric tensor is diagonal with all the diagonal elements being +1. This also applies to plane surfaces which are 2D flat spaces that can be coordinated by 2D orthonormal Cartesian systems. Another example is the 4D Minkowski space-time manifold associated with the mechanics of Lorentz transformations whose metric is diagonal with elements of ±1 (see Eq. 241). When all the diagonal elements of the metric tensor of a flat space are +1, the space and the coordinate system may be described as homogeneous (see § 2.2.3). All 1D spaces are Euclidean and hence they cannot be curved intrinsically, so twisted curves are curved only when viewed externally from the embedding space which they reside in, e.g. the 2D space of a surface curve or the 3D space of a space curve. This is because any curve can be mapped isometrically to a straight line where both are naturally parameterized by arc length. An example of curved space is the 2D surface of a sphere or an ellipsoid since there is no possibility of coordinating these spaces with valid 2D coordinate systems that satisfy the above criterion.

A curved space may have constant curvature all over the space, or have variable curvature and hence the curvature is position dependent. An example of a space of constant curvature is the surface of a sphere of radius R whose curvature (i.e. Riemannian curvature) is $\frac{1}{R^2}$ at each point of the surface. Torus and ellipsoid are simple examples of 2D spaces with variable curvature. Schur theorem related to nD spaces ($n > 2$) of constant curvature states that: if the Riemann-Christoffel curvature tensor (see § 7.2.1) at each point of a space is a function of the coordinates only, then the curvature is constant all over the space. Schur theorem may also be stated as: the Riemannian curvature is constant over an isotropic region of an nD ($n > 2$) Riemannian space.

A necessary and sufficient condition for an nD space to be intrinsically flat is that the Riemann-Christoffel curvature tensor of the space vanishes identically. Hence, cylinders are intrinsically flat, since their Riemann-Christoffel curvature tensor vanishes identically, although they are curved as seen extrinsically from the embedding 3D space. On the other hand, planes are intrinsically and extrinsically flat. In brief, a space is intrinsically flat *iff* the Riemann-Christoffel curvature tensor vanishes identically over the space, and it is extrinsically (as well as intrinsically) flat *iff* the curvature tensor vanishes identically over the whole space. This is because the Riemann-Christoffel curvature tensor characterizes the space curvature from an intrinsic perspective while the curvature tensor characterizes the space curvature from an extrinsic perspective.

As indicated above, the geometry of curved spaces is usually described as the Riemannian geometry. One approach for investigating the Riemannian geometry of a curved manifold is to embed the manifold in a Euclidean space of higher dimensionality and inspect the properties of the manifold from this perspective. This approach is largely followed, for example, in the differential geometry of surfaces where the geometry of curved 2D spaces (twisted surfaces) is investigated by immersing the surfaces in a 3D Euclidean space and examining their properties as viewed from this external enveloping 3D space. Such an external view is necessary for examining the extrinsic geometry of the space but not its intrinsic geometry. A similar approach may also be followed in the investigation of surface and space curves.

2.2 Coordinate Systems

In simple terms, a coordinate system is a mathematical device, essentially of geometric nature, used by an observer to identify the location of points and objects and describe events in generalized space which may include space-time. In tensor calculus, a coordinate system is needed to define non-scalar tensors in a specific form and identify their components in reference to the basis set of the system. An nD space requires a coordinate system with n mutually independent variable coordinates to be fully described so that any point in the space can be uniquely identified by the coordinate system. We note that the coordinates are generally real quantities although this may not apply in some cases (see § 2.2.3).

As we will see in § 2.4, coordinate systems of 3D spaces are characterized by having coordinate curves and coordinate surfaces where the coordinate curves occur at the intersection of the coordinate surfaces.[5] The coordinate curves represent the curves along which exactly one coordinate varies while the other coordinates are held constant. Conversely, the coordinate surfaces represent the surfaces over which exactly one coordinate is held constant while the other coordinates vary. At any point P in a 3D space coordinated by a 3D coordinate system, we have 3 independent coordinate curves and 3 independent coordinate surfaces passing through P. The 3 coordinate curves uniquely identify the set of 3 mutually independent covariant basis vectors at P. Similarly, the 3 coordinate surfaces uniquely identify the set of 3 mutually independent contravariant basis vectors at P. Further details about this issue will follow in § 2.4.

There are many types and categories of coordinate system; some of which will be briefly investigated in the following subsections. The most commonly used coordinate systems are: Cartesian, cylindrical and spherical. The most universal type of coordinate system is the general coordinate system which can include any type (rectilinear, curvilinear, orthogonal, etc.). A subset of the general coordinate system is the orthogonal coordinate system which is characterized by having mutually perpendicular coordinate curves, as well as mutually

---

[5] In fact, these concepts can be generalized to nD spaces by generalizing the concepts of curves and surfaces. However, the main interest here and in the forthcoming sections is 3D spaces.

perpendicular coordinate surfaces, at each point in the region of space over which the system is defined and hence its basis vectors, whether covariant or contravariant, are mutually perpendicular.

The coordinates of a system can have the same physical dimension or different physical dimensions. An example of the first is the Cartesian coordinate system, which is usually identified by (x, y, z), where all the coordinates have the dimension of length, while examples of the second include the cylindrical and spherical systems, which are usually identified by (ρ, ϕ, z) and (r, θ, ϕ) respectively, where some coordinates, like ρ and r, have the dimension of length while other coordinates, like ϕ and θ, are dimensionless. We also note that the physical dimensions of the components and basis vectors of the covariant and contravariant forms of a tensor are generally different.

In the following subsections we outline a number of general types and categories of coordinate systems based on different classifying criteria. These categories are generally overlapping and may not be exhaustive in their domain.

2.2.1 Rectilinear and Curvilinear Coordinate Systems

Rectilinear coordinate systems are characterized by the property that all their coordinate curves are straight lines and all their coordinate surfaces are planes, while curvilinear coordinate systems are characterized by the property that at least some of their coordinate curves are not straight lines and some of their coordinate surfaces are not planes (see § 2.4). Consequently, the basis vectors of rectilinear systems are constant while the basis vectors of curvilinear systems are variable in general since their direction or/and magnitude depend on the position in the space and hence they are coordinate dependent.

Rectilinear coordinate systems can be rectangular (or orthogonal) when their coordinate curves, as well as their coordinate surfaces, are mutually orthogonal such as the well known rectangular Cartesian system. They can also be oblique when at least some of their coordinate curves and coordinate surfaces do not satisfy this condition. Figure 2 is a simple graphic illustration of rectangular and oblique rectilinear coordinate systems in a 3D space. Similarly, curvilinear coordinate systems can be orthogonal, when the vectors in their covariant or contravariant basis set are mutually orthogonal at each point in the space, and can be non-orthogonal when this condition is not met. Rectilinear coordinate systems may also be labeled as affine or linear coordinate systems although the terminology is not universal and hence these labels may be used differently.

As stated above, curvilinear coordinate systems are characterized by the property that at least some of their coordinate curves are not straight lines and some of their coordinate surfaces are not planes (see Figures 3 and 10). This means that some (but not all) of the coordinate curves of curvilinear coordinate systems can be straight lines and some (but not all) of their coordinate surfaces can be planes. This is the case in the cylindrical and spherical coordinate systems as we will see next. Also, the coordinate curves of curvilinear coordinate systems may be regularly shaped curves such as circles and may be irregularly shaped and hence they are generalized twisted curves. Similarly, the coordinate surfaces of curvilinear systems may be regularly shaped surfaces such as spheres and may

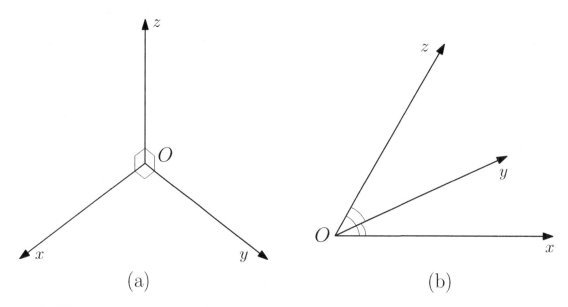

Figure 2: The two main types of rectilinear coordinate systems in 3D spaces: (a) rectangular and (b) oblique.

be irregularly shaped.

Prominent examples of curvilinear coordinate systems are the cylindrical and spherical systems of 3D spaces. All the coordinate curves and coordinate surfaces of these systems are regularly shaped. As we will see (refer to § 2.4), in the cylindrical coordinate systems the ρ, ϕ, z coordinate curves are straight lines, circles and straight lines respectively, while the ρ, ϕ, z coordinate surfaces are cylinders, semi-planes and planes respectively. Similarly, in the spherical coordinate systems the r, θ, ϕ coordinate curves are straight lines, semi-circles and circles respectively, while the r, θ, ϕ coordinate surfaces are spheres, cones and semi-planes respectively. We note that an admissible coordinate transformation from a rectilinear system defines another rectilinear system if the transformation is linear, and defines a curvilinear system if the transformation is nonlinear.

2.2.2 Orthogonal Coordinate Systems

The characteristic feature of orthogonal coordinate systems, whether rectilinear or curvilinear, is that their coordinate curves, as well as their coordinate surfaces, are mutually perpendicular at each point in their space. Hence, the vectors of their covariant basis set and the vectors of their contravariant basis set are mutually orthogonal. As a result, the corresponding covariant and contravariant basis vectors in orthogonal coordinate systems have the same direction and therefore if the vectors of these basis sets are normalized they will be identical, i.e. the normalized covariant and the normalized contravariant basis vector sets are the same.

Prominent examples of orthogonal coordinate systems are rectangular Cartesian, cylindrical and spherical systems of 3D spaces (refer to Figure 4). A necessary and sufficient condition for a coordinate system to be orthogonal is that its metric tensor is diagonal.

2.2.3 Homogeneous Coordinate Systems

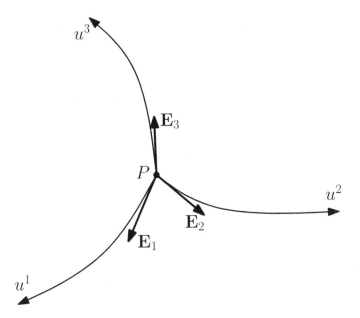

Figure 3: General curvilinear coordinate system in a 3D space and its covariant basis vectors $\mathbf{E}_1, \mathbf{E}_2$ and $\mathbf{E}_3$ (see § 2.6) as tangents to the shown coordinate curves at a particular point of the space P, where u^1, u^2 and u^3 represent general coordinates.

This can be inferred from the definition of the components of the metric tensor as dot products of the basis vectors (see Eq. 48) since the dot product involving two different vectors (i.e. $\mathbf{E}_i \cdot \mathbf{E}_j$ or $\mathbf{E}^i \cdot \mathbf{E}^j$ with $i \neq j$) will vanish if the basis vectors, whether covariant or contravariant, are mutually perpendicular.

2.2.3 Homogeneous Coordinate Systems

When all the diagonal elements of a diagonal metric tensor of a flat space are $+1$, the coordinate system is described as homogeneous. In this case the length of line element ds of Eq. 33 becomes:

$$(ds)^2 = du^i du^i \tag{34}$$

An example of homogeneous coordinate systems is the orthonormal Cartesian system of a 3D Euclidean space (Figure 4 a). A homogeneous coordinate system can be transformed to another homogeneous coordinate system only by linear transformations. Moreover, any coordinate system obtained from a homogeneous coordinate system by an orthogonal transformation (see § 2.3.3) is also homogeneous. As a consequence of the last statements, infinitely many homogeneous coordinate systems can be constructed in any flat space.

A coordinate system of a flat space can always be homogenized by allowing the coordinates to be imaginary. This is done by redefining the coordinates as:

$$U^i = \sqrt{\zeta_i} u^i \qquad \text{(no sum over } i\text{)} \tag{35}$$

where $\zeta_i = \pm 1$. The new coordinates U^i are real when $\zeta_i = 1$ and imaginary when $\zeta_i = -1$.

2.2.3 Homogeneous Coordinate Systems

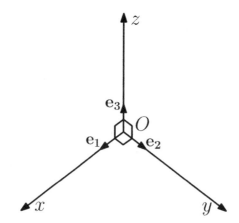

(a) Orthonormal Cartesian system and its orthonormal basis vectors $\mathbf{e}_1, \mathbf{e}_2, \mathbf{e}_3$

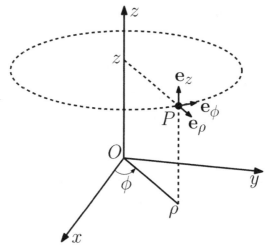

(b) Cylindrical coordinate system and its orthonormal basis vectors $\mathbf{e}_\rho, \mathbf{e}_\phi, \mathbf{e}_z$

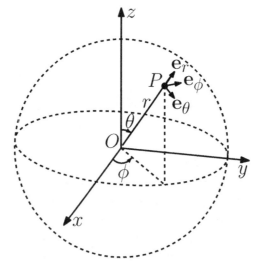

(c) Spherical coordinate system and its orthonormal basis vectors $\mathbf{e}_r, \mathbf{e}_\theta, \mathbf{e}_\phi$

Figure 4: The three prominent orthogonal coordinate systems in 3D spaces: (a) orthonormal Cartesian, (b) cylindrical and (c) spherical.

Consequently, the length of line element ds will be given by:

$$(ds)^2 = dU^i dU^i \qquad (36)$$

which is of the same form as Eq. 34. An example of a homogeneous coordinate system with some real and some imaginary coordinates is the coordinate system of a Minkowski 4D space-time of the mechanics of Lorentz transformations. We note that homogenization in the above sense is based on an extension of the concept of homogeneity and it is mainly based on the definition of the length of line element.

2.3 Transformations

In general terms, a transformation from an nD space to another nD space is a correlation that maps a point from the first space (original) to a point in the second space (transformed) where each point in the original and transformed spaces is identified by n independent coordinates. To distinguish between the two sets of coordinates in the two spaces, the coordinates of the points in the transformed space may be notated with barred symbols like $(\bar{u}^1, \bar{u}^2, \ldots, \bar{u}^n)$, while the coordinates of the points in the original space are notated with unbarred similar symbols like $(u^1, u^2, \ldots, u^n)$. Under certain conditions, which will be clarified later, such a transformation is unique and hence an inverse transformation from the transformed space to the original space is also defined.

Mathematically, each one of the direct and inverse transformations can be regarded as a mathematical correlation expressed by a set of equations in which each coordinate in one space is considered as a function of the coordinates in the other space. Hence, the transformations between the two sets of coordinates in the two spaces can be expressed mathematically in a generic form by the following two sets of independent relations:

$$\bar{u}^i = \bar{u}^i(u^1, u^2, \ldots, u^n) \qquad\qquad u^i = u^i(\bar{u}^1, \bar{u}^2, \ldots, \bar{u}^n) \qquad (37)$$

where $i = 1, 2, \ldots, n$ with n being the space dimension. The independence of the above relations is guaranteed *iff* the Jacobian of the transformation does not vanish at any point in the space (refer to the following paragraphs about the Jacobian).

An alternative to the latter view of considering the transformation as a mapping between two different spaces is to view it as a correlation relating the same point in the same space but observed from two different coordinate systems which are subject to a similar transformation. The following will be largely based on the latter view although we usually adopt the one which is more convenient in the particular context. As far as the notation is concerned, there is no fundamental difference between the barred and unbarred systems and hence the notation can be interchanged. We also note that the transformations considered here, and in the present book in general, are between two spaces or coordinate systems of equal dimensions and hence we do not consider transformations between spaces or coordinate systems of different dimensions. Consequently, the Jacobian matrix of the transformation is always square and its determinant (i.e. the Jacobian) is defined. As indicated earlier, if the mapping from an original rectangular Cartesian system is linear,

2.3 Transformations

the coordinate system obtained from such a transformation is called affine or rectilinear. Coordinate systems which are not affine are described as curvilinear although the terminology may differ between the authors.

The following $n \times n$ matrix of n^2 partial derivatives of the unbarred coordinates with respect to the barred coordinates, where n is the space dimension, is called the "Jacobian matrix" of the transformation between the unbarred and barred systems:

$$\mathbf{J} = \begin{bmatrix} \frac{\partial u^1}{\partial \bar{u}^1} & \frac{\partial u^1}{\partial \bar{u}^2} & \cdots & \frac{\partial u^1}{\partial \bar{u}^n} \\ \frac{\partial u^2}{\partial \bar{u}^1} & \frac{\partial u^2}{\partial \bar{u}^2} & \cdots & \frac{\partial u^2}{\partial \bar{u}^n} \\ \vdots & \vdots & \ddots & \vdots \\ \frac{\partial u^n}{\partial \bar{u}^1} & \frac{\partial u^n}{\partial \bar{u}^2} & \cdots & \frac{\partial u^n}{\partial \bar{u}^n} \end{bmatrix} \qquad (38)$$

while its determinant:

$$J = \det(\mathbf{J}) \qquad (39)$$

is called the "Jacobian" of the transformation where the indexed u and $\bar{u}$ are the coordinates in the unbarred and barred coordinate systems of the nD space. The Jacobian array (whether matrix or determinant) contains all the possible n^2 partial derivatives made of the different combinations of the u and $\bar{u}$ indices where the pattern of the indices in this array is simple, that is the indices of u in the numerator provide the indices for the rows while the indices of $\bar{u}$ in the denominator provide the indices for the columns. This labeling scheme may be interchanged which is equivalent to taking the transpose of the array. As it is well known, the Jacobian will not change by this transposition since the determinant of a matrix is the same as the determinant of its transpose, i.e. $\det(\mathbf{A}) = \det(\mathbf{A}^T)$.

We note that all coordinate transformations in the present book are continuous, single valued and invertible. We also note that "barred" and "unbarred" in the definition of Jacobian should be understood in a general sense not just as two labels since the Jacobian is not restricted to transformations between two systems of the same type but labeled as barred and unbarred. In fact the two coordinate systems can be fundamentally different in nature such as orthonormal Cartesian and general curvilinear. The Jacobian matrix and determinant represent any transformation by the above partial derivative array system between two coordinate systems defined by two different sets of coordinate variables not necessarily as barred and unbarred. The objective of defining the Jacobian as between unbarred and barred systems is simplicity and generality.

The transformation from the unbarred coordinate system to the barred coordinate system is bijective[6] *iff* $J \neq 0$ at any point in the transformed region of the space. In this case, the inverse transformation from the barred to the unbarred system is also defined and bijective and is represented by the inverse of the Jacobian matrix, that is:

$$\bar{\mathbf{J}} = \mathbf{J}^{-1} \qquad (40)$$

Consequently, the Jacobian of the inverse transformation, being the determinant of the inverse Jacobian matrix, is the reciprocal of the Jacobian of the original transformation,

---

[6] Bijective means the mapping is injective (one-to-one) and surjective (onto).

2.3 Transformations

that is:
$$\bar{J} = \frac{1}{J} \qquad (41)$$

As we remarked, there is no fundamental notational difference between the barred and unbarred systems and hence the labeling is rather arbitrary and can be interchanged. Therefore, the Jacobian may be notated as unbarred over barred or the other way around. The essence is that the Jacobian usually represents the transformation from an original system to another system while its inverse represents the opposite transformation although even this is not generally respected in the literature of mathematics and hence "Jacobian" my be used to label the opposite transformation. Yes, in a specific context when one of these is labeled as the Jacobian, the other one should be labeled as the inverse Jacobian to distinguish between the two opposite Jacobians and their corresponding transformations. However, there may also be practical aspects for choosing which is the Jacobian and which is the inverse since it is easier sometimes to compute one of these than the other and hence we start by computing the easier as the Jacobian (whether from original to transformed or the other way) followed by obtaining the reciprocal as the inverse Jacobian. Anyway, in this book we generally use "Jacobian" flexibly where the context determines the nature of the transformation.[7]

An admissible (or permissible or allowed) coordinate transformation may be defined generically as a mapping represented by a sufficiently differentiable set of equations plus being invertible by having a non-vanishing Jacobian ($J \neq 0$). More technically, a coordinate transformation is commonly described as admissible *iff* the transformation is bijective with non-vanishing Jacobian and the transformation function is of class C^2.[8] We note that the C^n continuity condition means that the function and all its first n partial derivatives do exist and are continuous in their domain. Also, some authors may impose a weaker continuity condition of being of class C^1.

An object that does not change by admissible coordinate transformations is described as "invariant" such as a true scalar (see § 3.1.2) which is characterized by its sign and magnitude and a true vector which is characterized by its magnitude and direction in space. Similarly, an invariant property of an object or a manifold is a property that is independent of admissible coordinate transformations such as being form invariant which characterizes tensors or being flat which characterize spaces under certain types of transformation. It should be noted that an invariant object or property may be invariant with respect to certain types of transformation but not with respect to other types of transformation and hence the term may be used generically where the context is taken into consideration for

---

[7] We may get rid of all these complications by using the term "Jacobian" to represent a system of partial derivatives whose objective is to transform from one coordinate system to another (regardless of anything else like system notation or being original and secondary) and hence both the "Jacobian" and its inverse are Jacobians in this general sense; the two will be distinguished from each other by stating from which system to which system the Jacobian transforms. Hence, we can say legitimately and with no ambiguity: the Jacobian from A to B and the Jacobian from B to A. These may be labeled as $J(A \to B)$ and $J(B \to A)$ with similar notation for the Jacobian matrix. This, we believe, can resolve all these issues and avoid confusion.

[8] The meaning of "admissible coordinate transformation" may vary depending on the context.

sensible interpretation.

A product or composition of space or coordinate transformations is a succession of transformations where the output of one transformation is taken as the input to the next transformation. For example, a series of m transformations labeled as T_i where $i = 1, 2, \cdots, m$ may be applied sequentially onto a mathematical object $\mathbb{O}$. This operation can be expressed mathematically by the following composite transformation T_c:

$$T_c(\mathbb{O}) = T_m T_{m-1} \cdots T_2 T_1(\mathbb{O}) \tag{42}$$

where the output of T_1 in this notation is taken as the input to T_2 and so forth until the output of T_{m-1} is fed as an input to T_m in the end to produce the final output of T_c. In such cases, the Jacobian of the product is the product of the Jacobians of the individual transformations of which the product is made, that is:

$$J_c = J_m J_{m-1} \cdots J_2 J_1 \tag{43}$$

where J_i in this notation is the Jacobian of the T_i transformation and J_c is the Jacobian of T_c.

The collection of all admissible coordinate transformations with non-vanishing Jacobian form a group. This means that they satisfy the properties of closure, associativity, identity and inverse. Hence, any convenient coordinate system can be chosen as the point of entry since other systems can be reached, if needed, through the set of admissible transformations. This is one of the cornerstones of building invariant physical theories which are independent of the subjective choice of coordinate systems and reference frames. We remark that transformation of coordinates is not a commutative operation and hence the result of two successive transformations may depend on the order of these transformations. This is demonstrated in Figure 5 where the composition of two rotations results in different outcomes depending on the order of the rotations.

As there are essentially two different types of basis vectors, namely tangent vectors of covariant nature and gradient vectors of contravariant nature (see § 2.6 and 3.1.1), there are two main types of non-scalar tensors: contravariant tensors and covariant tensors which are based on the type of the employed basis vectors of the given coordinate system. Tensors of mixed type employ in their definition mixed basis vectors of the opposite type to the corresponding indices of their components. As we will see, the transformation between these different types is facilitated by the metric tensor of the given coordinate system (refer to § 4.5).

In the following subsections, we briefly describe a number of types and categories of coordinate transformations.

2.3.1 Proper and Improper Transformations

Coordinate transformations are described as "proper" when they preserve the handedness (right- or left-handed) of the coordinate system and "improper" when they reverse the handedness. Improper transformations involve an odd number of coordinate axes inversions in the origin of coordinates. Inversion of axes may be called improper rotation while

2.3.2 Active and Passive Transformations 33

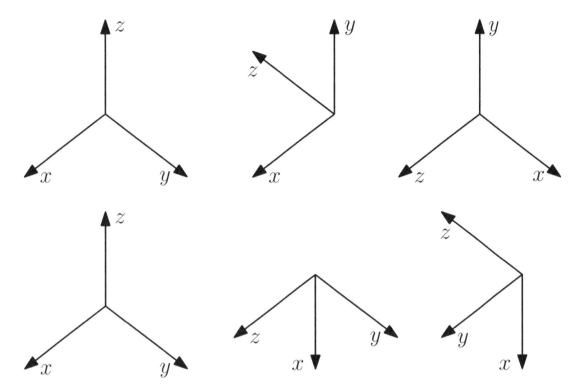

Figure 5: The composite transformations $R_y R_x$ (top) and $R_x R_y$ (bottom) where R_x and R_y represent clockwise rotation of $\frac{\pi}{2}$ around the positive x and y axes respectively.

ordinary rotation is described as proper rotation. Figure 6 illustrates proper and improper coordinate transformations of a rectangular Cartesian coordinate system in a 3D space.

2.3.2 Active and Passive Transformations

Transformations can be active, when they change the state of the observed object such as rotating the object in the space, or passive when they are based on keeping the state of the object and changing the state of the coordinate system which the object is observed from. In brief, the subject of an active transformation is the object while the subject of a passive transformation is the coordinate system.

2.3.3 Orthogonal Transformations

An orthogonal coordinate transformation consists of a combination of translation, rotation and reflection of axes. The Jacobian of orthogonal transformations is unity, that is $J = \pm 1$.[9] The orthogonal transformation is described as positive *iff* $J = +1$ and neg-

---

[9] This condition should apply even when the transformation includes a translation since the added constants that represent the translation in the transformation equations will vanish in the Jacobian matrix. However, there seems to be a different convention that excludes translation to be part of orthogonal transformations. There also seems to be another convention which restricts orthogonal transformations to translation and rotation.

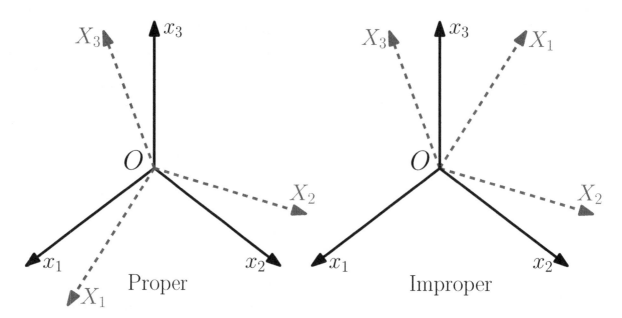

Figure 6: Proper and improper transformations of a rectangular Cartesian coordinate system in a 3D space where the former is achieved by a rotation of the coordinate system while the latter is achieved by a rotation followed by a reflection of the first axis in the origin of coordinates. The transformed systems are shown as dashed and labeled with upper case letters while the original system is shown as solid and labeled with lower case letters.

ative *iff* $J = -1$. Positive orthogonal transformations consist solely of translation and rotation (possibly trivial ones as in the case of the identity transformation) while negative orthogonal transformations include reflection, by applying an odd number of axes reversal, as well. Positive transformations can be decomposed into an infinite number of continuously varying infinitesimal positive transformations each one of which imitates an identity transformation. Such a decomposition is not possible in the case of negative orthogonal transformations because the shift from the identity transformation to reflection is impossible by a continuous process.

2.3.4 Linear and Nonlinear Transformations

The characteristic property of linear transformations is that they maintain scalar multiplication and algebraic addition, while nonlinear transformations do not. Hence, if T is a linear transformation then we have:

$$T(a\mathbf{A} \pm b\mathbf{B}) = a\,T(\mathbf{A}) \pm b\,T(\mathbf{B}) \qquad (44)$$

where $\mathbf{A}$ and $\mathbf{B}$ are mathematical objects to be transformed by T and a and b are scalars. As indicated earlier, an admissible coordinate transformation from a rectilinear system defines another rectilinear system if the transformation is linear, and defines a curvilinear system if the transformation is nonlinear.

2.4 Coordinate Curves and Coordinate Surfaces

As seen earlier, coordinate systems of 3D spaces are characterized by having coordinate curves and coordinate surfaces where the coordinate curves represent the curves of mutual intersection of the coordinate surfaces.[10] These coordinate curves and coordinate surfaces play a crucial role in the formulation and development of the mathematical structures of the coordinated space. The coordinate curves represent the curves along which exactly one coordinate varies while the other coordinates are held constant. Conversely, the coordinate surfaces represent the surfaces over which all coordinates vary except one which is held constant. In brief, the ith coordinate curve is the curve along which only the ith coordinate varies while the ith coordinate surface is the surface over which only the ith coordinate is constant.

For example, in a 3D Cartesian system identified by the coordinates (x, y, z) the curve $\mathbf{r}(x) = (x, c_2, c_3)$, where c_2 and c_3 are real constants, is an x coordinate curve since x varies while y and z are held constant, and the surface $\mathbf{r}(x, z) = (x, c_2, z)$ is a y coordinate surface since x and z vary while y is held constant. Similarly, in a cylindrical coordinate system identified by the coordinates (ρ, ϕ, z) the curve $\mathbf{r}(\phi) = (c_1, \phi, c_3)$, where c_1 is a real constant, is a ϕ coordinate curve since ρ and z are held constant while ϕ varies. Likewise, in a spherical coordinate system identified by the coordinates (r, θ, ϕ) the surface $\mathbf{r}(\theta, \phi) = (c_1, \theta, \phi)$ is an r coordinate surface since r is held constant while θ and ϕ vary.

As stated before, coordinate curves represent the curves of mutual intersection of coordinate surfaces. This should be obvious since along the intersection curve of two coordinate surfaces, where on each one of these surfaces one coordinate is held constant, two coordinates will be constant and hence only the third coordinate can vary. Hence, in a 3D Cartesian coordinate system the x coordinate curves occur at the intersection of the y and z coordinate surfaces, the y coordinate curves occur at the intersection of the x and z coordinate surfaces, and the z coordinate curves occur at the intersection of the x and y coordinate surfaces (refer to Figure 7). Similarly, in a cylindrical coordinate system the ρ, ϕ and z coordinate curves occur at the intersection of the ϕ, z, the ρ, z and the ρ, ϕ coordinate surfaces respectively (refer to Figure 8). Likewise, in a spherical coordinate system the r, θ and ϕ coordinate curves occur at the intersection of the θ, ϕ, the r, ϕ and the r, θ coordinate surfaces respectively (refer to Figure 9).

In this context, we note that the three types of coordinate surface of a Cartesian system, or in fact any rectilinear coordinate system (see § 2.2.1), are planes and hence the three types of coordinate curves are straight lines (refer to Figure 7). As for cylindrical coordinate systems, the ρ coordinate surfaces are cylinders (and this maybe the reason behind labeling them as "cylindrical"), the ϕ coordinate surfaces are semi-planes and the z coordinate surfaces are planes. Hence, the ρ, ϕ and z coordinate curves are straight lines, circles, and straight lines respectively (refer to Figure 8). Regarding spherical coordinate systems, the r coordinate surfaces are spheres, the θ coordinate surfaces are cones and the ϕ coordinate surfaces are semi-planes. Hence, the r, θ and ϕ coordinate curves are

---

[10] Our focus here is 3D spaces although these concepts can be generalized to nD spaces as indicated earlier.

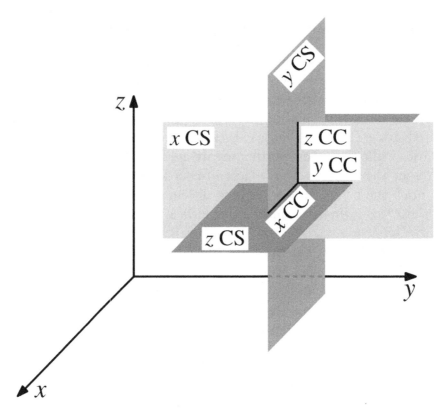

Figure 7: Coordinate curves (CC) and coordinate surfaces (CS) in a 3D Cartesian coordinate system.

straight lines, semi-circles, and circles respectively (refer to Figure 9).

The coordinate surfaces of general curvilinear coordinate systems can vary in general and hence they are not represented by particular geometric shapes such as planes and spheres. As a result, the coordinate curves of these systems are general space curves which may not have regular geometric shapes such as straight line or circle. As stated earlier, orthogonal coordinate systems are characterized by having coordinate surfaces which are mutually orthogonal at each point of the space, and consequently their coordinate curves are also mutually orthogonal at each point of the space. Finally, we remark that the transformation relations of Eq. 37 are used in defining the set of coordinate surfaces and coordinate curves.

2.5 Scale Factors

Scale factors of a coordinate system are those factors which are required to multiply the coordinate differentials to obtain the distances traversed during a change in the coordinate of that magnitude. The scale factors are symbolized with $h_1, h_2, \ldots, h_n$ where this notation is usually used in orthogonal coordinate systems. For example, in the plane polar coordinate system represented by the coordinates (ρ, ϕ), the scale factor of the second coordinate ϕ is ρ because ρ is the factor used to multiply the differential of the polar angle $d\phi$ to obtain the distance L traversed by a change of magnitude $d\phi$ in the polar angle,

2.5 Scale Factors

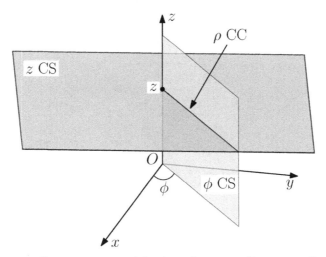

(a) ρ coordinate curve with ϕ and z coordinate surfaces

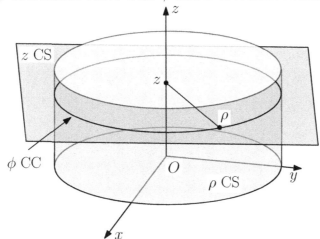

(b) ϕ coordinate curve with ρ and z coordinate surfaces

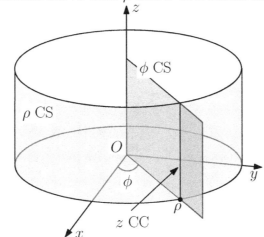

(c) z coordinate curve with ρ and ϕ coordinate surfaces

Figure 8: Coordinate curves (CC) and coordinate surfaces (CS) in cylindrical coordinate systems.

2.5 Scale Factors

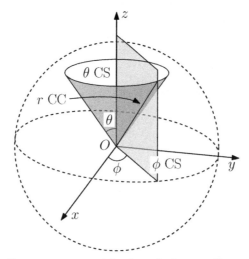

(a) r coordinate curve with θ and ϕ coordinate surfaces

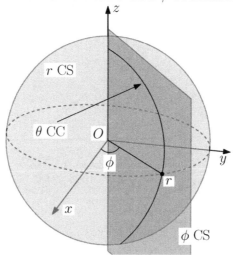

(b) θ coordinate curve with r and ϕ coordinate surfaces

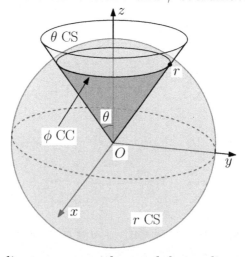

(c) ϕ coordinate curve with r and θ coordinate surfaces

Figure 9: Coordinate curves (CC) and coordinate surfaces (CS) in spherical coordinate systems.

Table 1: The scale factors (h_1, h_2, h_3) for the three most commonly used orthogonal coordinate systems in 3D spaces: orthonormal Cartesian, cylindrical and spherical. The squares of these entries and the reciprocals of these squares give the diagonal elements of the covariant and contravariant metric tensors, g_{ij} and g^{ij}, respectively of these systems (see Eqs. 238-240).

	Cartesian (x, y, z)	Cylindrical (ρ, ϕ, z)	Spherical (r, θ, ϕ)
h_1	1	1	1
h_2	1	ρ	r
h_3	1	1	$r \sin\theta$

that is: $L = \rho \, d\phi$.

The scale factors are also used for other purposes such as normalizing the basis vectors and defining the components in the physical representation of vectors and tensors (refer to § 2.6 and 3.3). They are also used in the analytical expressions for length, area and volume in orthogonal coordinate systems, as described in § 4.6.8, 4.6.9 and 4.6.10. The scale factors for the Cartesian, cylindrical and spherical coordinate systems of 3D spaces are given in Table 1.

2.6 Basis Vectors and Their Relation to the Metric and Jacobian

The set of basis vectors in a given manifold plays a pivotal role in the theoretical construction of the geometry of the manifold, where these vectors are used in the definition and construction of essential concepts and objects such as the metric tensor of the space. The basis sets are defined at each regular point in the space and hence the vectors in the basis sets vary in general from one point to another, i.e. they are coordinate dependent.

The vectors providing the basis set for a coordinate system, which are not necessarily of unit length or mutually orthogonal, can be of covariant type or contravariant type. The covariant basis vectors are defined as the tangent vectors to the coordinate curves, while the contravariant basis vectors are defined as the gradient of the space coordinates and hence they are perpendicular to the coordinate surfaces (refer to Figure 10). Formally, the covariant and contravariant basis vectors are defined respectively by:

$$\mathbf{E}_i = \frac{\partial \mathbf{r}}{\partial u^i} \qquad \mathbf{E}^i = \nabla u^i \qquad (45)$$

where $\mathbf{r}$ is the position vector in Cartesian coordinates $(x^1, \ldots, x^n)$, u^i represents general coordinates, n is the space dimension and $i = 1, \cdots, n$. As indicated above, the covariant and contravariant basis sets, $\mathbf{E}_i$ and $\mathbf{E}^i$, in general coordinate systems are functions of coordinates, that is:

$$\mathbf{E}_i = \mathbf{E}_i\left(u^1, \ldots, u^n\right) \qquad \mathbf{E}^i = \mathbf{E}^i\left(u^1, \ldots, u^n\right) \qquad (46)$$

Hence, the definitions of Eq. 45 apply to each individual point in the space where the coordinate curves and coordinate surfaces belong to that particular point. For example, at any particular point P in a 3D space with a valid coordinate system we have three mutually independent coordinate curves and three mutually independent coordinate surfaces and hence we should have three mutually independent covariant basis vectors and three mutually independent contravariant basis vectors at P.

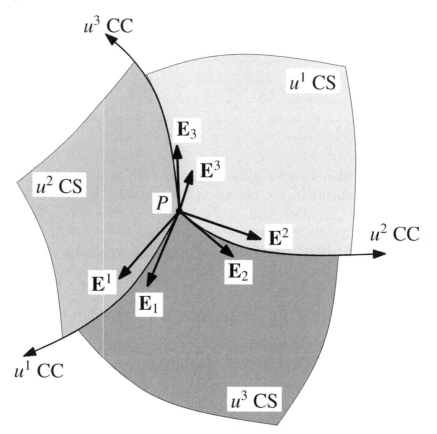

Figure 10: The covariant and contravariant basis vectors of a general curvilinear coordinate system and the associated coordinate curves (CC) and coordinate surfaces (CS) at a given point P in a 3D space.

Like other vectors, the covariant and contravariant basis vectors of a given coordinate system are related to each other, through the metric tensor of the system, by the following relations:

$$\mathbf{E}_i = g_{ij}\mathbf{E}^j \qquad\qquad \mathbf{E}^i = g^{ij}\mathbf{E}_j \qquad (47)$$

where the metric tensor in its covariant form g_{ij} and contravariant form g^{ij} works as an index shifting operator to lower and raise the indices and hence change the variance type of the basis vectors. Thus, the transformation between the covariant basis set and the contravariant basis set of a particular coordinate system is facilitated by the metric tensor of that system.

The basis vectors in their covariant and contravariant forms are related to the components of the metric tensor in its covariant and contravariant forms by the following relations:

$$\mathbf{E}_i \cdot \mathbf{E}_j = g_{ij} \qquad\qquad \mathbf{E}^i \cdot \mathbf{E}^j = g^{ij} \qquad (48)$$

2.6 Basis Vectors and Their Relation to the Metric and Jacobian

The covariant and contravariant basis vectors are reciprocal basis systems, and hence we have:

$$\mathbf{E}_i \cdot \mathbf{E}^j = \delta_i^{\ j} \qquad \mathbf{E}^i \cdot \mathbf{E}_j = \delta^i_{\ j} \qquad (49)$$

where the indexed δ is the Kronecker delta tensor (refer to § 4.1). As we will see in § 4.5, the Kronecker delta tensor represents the metric tensor in its mixed form.

In a 3D space with a right handed coordinate system the two sets of basis vectors are linked by the following relations:

$$\mathbf{E}_1 = \frac{\mathbf{E}^2 \times \mathbf{E}^3}{\mathbf{E}^1 \cdot (\mathbf{E}^2 \times \mathbf{E}^3)}, \qquad \mathbf{E}_2 = \frac{\mathbf{E}^3 \times \mathbf{E}^1}{\mathbf{E}^1 \cdot (\mathbf{E}^2 \times \mathbf{E}^3)}, \qquad \mathbf{E}_3 = \frac{\mathbf{E}^1 \times \mathbf{E}^2}{\mathbf{E}^1 \cdot (\mathbf{E}^2 \times \mathbf{E}^3)} \qquad (50)$$

$$\mathbf{E}^1 = \frac{\mathbf{E}_2 \times \mathbf{E}_3}{\mathbf{E}_1 \cdot (\mathbf{E}_2 \times \mathbf{E}_3)}, \qquad \mathbf{E}^2 = \frac{\mathbf{E}_3 \times \mathbf{E}_1}{\mathbf{E}_1 \cdot (\mathbf{E}_2 \times \mathbf{E}_3)}, \qquad \mathbf{E}^3 = \frac{\mathbf{E}_1 \times \mathbf{E}_2}{\mathbf{E}_1 \cdot (\mathbf{E}_2 \times \mathbf{E}_3)} \qquad (51)$$

The pattern in these relations is very simple, that is the covariant vectors are represented by contravariant vectors and vice versa; moreover, the indices in the numerators on both sides represent even permutations of $1, 2, 3$.

The relations in Eqs. 50 and 51 may be expressed in a more compact form as follows:

$$\mathbf{E}_i = \frac{\mathbf{E}^j \times \mathbf{E}^k}{\mathbf{E}^i \cdot (\mathbf{E}^j \times \mathbf{E}^k)} \qquad \mathbf{E}^i = \frac{\mathbf{E}_j \times \mathbf{E}_k}{\mathbf{E}_i \cdot (\mathbf{E}_j \times \mathbf{E}_k)} \qquad (52)$$

where i, j, k take respectively the values $1, 2, 3$ and the other two cyclic permutations (i.e. $2, 3, 1$ and $3, 1, 2$). It is worth noting that the magnitude of the scalar triple product $\mathbf{E}_i \cdot (\mathbf{E}_j \times \mathbf{E}_k)$ represents the volume of the parallelepiped formed by the vectors $\mathbf{E}_i$, $\mathbf{E}_j$ and $\mathbf{E}_k$ while the magnitude of the scalar triple product $\mathbf{E}^i \cdot (\mathbf{E}^j \times \mathbf{E}^k)$ represents its reciprocal. We also note that the scalar triple product is invariant to a cyclic permutation of the symbols of the three vectors involved, i.e. $\mathbf{E}_i \cdot (\mathbf{E}_j \times \mathbf{E}_k) = \mathbf{E}_k \cdot (\mathbf{E}_i \times \mathbf{E}_j) = \mathbf{E}_j \cdot (\mathbf{E}_k \times \mathbf{E}_i)$; in fact its magnitude is invariant even to non-cyclic permutations. Another important note is that the magnitudes of the basis vectors in orthogonal coordinate systems are related to the scale factors of the coordinate system by:

$$|\mathbf{E}_i| = h_i \qquad |\mathbf{E}^i| = \frac{1}{h_i} \qquad (53)$$

where h_i is the scale factor for the ith coordinate (see § 2.5).

Following an admissible coordinate transformation between unbarred and barred general coordinate systems, the basis vectors in these systems are related by the following transformation rules:

$$\mathbf{E}_i = \frac{\partial \bar{u}^j}{\partial u^i} \bar{\mathbf{E}}_j \qquad \bar{\mathbf{E}}_i = \frac{\partial u^j}{\partial \bar{u}^i} \mathbf{E}_j \qquad (54)$$

$$\mathbf{E}^i = \frac{\partial u^i}{\partial \bar{u}^j} \bar{\mathbf{E}}^j \qquad \bar{\mathbf{E}}^i = \frac{\partial \bar{u}^i}{\partial u^j} \mathbf{E}^j \qquad (55)$$

where the indexed u and $\bar{u}$ represent the coordinates in the unbarred and barred systems respectively. The transformation rules for the components can be easily deduced from

2.6 Basis Vectors and Their Relation to the Metric and Jacobian

the above rules. For example, for a vector **A** which can be represented covariantly and contravariantly in the unbarred and barred systems as:

$$\mathbf{A} = \mathbf{E}^i A_i = \bar{\mathbf{E}}^i \bar{A}_i \tag{56}$$

$$\mathbf{A} = \mathbf{E}_i A^i = \bar{\mathbf{E}}_i \bar{A}^i \tag{57}$$

the transformation equations of its components between the two systems are given respectively by:

$$A_i = \frac{\partial \bar{u}^j}{\partial u^i} \bar{A}_j \qquad\qquad \bar{A}_i = \frac{\partial u^j}{\partial \bar{u}^i} A_j \tag{58}$$

$$A^i = \frac{\partial u^i}{\partial \bar{u}^j} \bar{A}^j \qquad\qquad \bar{A}^i = \frac{\partial \bar{u}^i}{\partial u^j} A^j \tag{59}$$

These transformation rules can be easily extended to higher rank tensors of different variance types (see § 3.1.1).

For a 3D manifold with a right handed coordinate system, we have:

$$\mathbf{E}_1 \cdot (\mathbf{E}_2 \times \mathbf{E}_3) = \sqrt{g} \qquad\qquad \mathbf{E}^1 \cdot \left(\mathbf{E}^2 \times \mathbf{E}^3\right) = \frac{1}{\sqrt{g}} \tag{60}$$

where g is the determinant of the covariant metric tensor, i.e.

$$g = \det(g_{ij}) = |g_{ij}| \tag{61}$$

Because $\mathbf{E}_i \cdot \mathbf{E}_j = g_{ij}$ (Eq. 48) we have:[11]

$$\mathbf{J}^T \mathbf{J} = [g_{ij}] \tag{62}$$

where **J** is the Jacobian matrix that transforms between Cartesian and general coordinates, the superscript T represents matrix transposition, $[g_{ij}]$ is the matrix representing the covariant metric tensor and the product on the left is a matrix product as defined in linear algebra which is equivalent to a dot product in tensor algebra. On taking the determinant of both sides of Eq. 62, the relation between the Jacobian and the determinant of the metric tensor is obtained, that is:

$$g = J^2 \tag{63}$$

where J is the Jacobian of the transformation. As it should be known, the determinant of a product is equal to the product of the determinants, that is:

$$\det(\mathbf{J}^T \mathbf{J}) = \det(\mathbf{J}^T) \det(\mathbf{J}) \tag{64}$$

---

[11] For example, for the transformation between Cartesian and general coordinate systems in 3D spaces we have:

$$\mathbf{J}^T = \begin{bmatrix} \frac{\partial x^1}{\partial u^1} & \frac{\partial x^2}{\partial u^1} & \frac{\partial x^3}{\partial u^1} \\ \frac{\partial x^1}{\partial u^2} & \frac{\partial x^2}{\partial u^2} & \frac{\partial x^3}{\partial u^2} \\ \frac{\partial x^1}{\partial u^3} & \frac{\partial x^2}{\partial u^3} & \frac{\partial x^3}{\partial u^3} \end{bmatrix} \qquad \mathbf{J} = \begin{bmatrix} \frac{\partial x^1}{\partial u^1} & \frac{\partial x^1}{\partial u^2} & \frac{\partial x^1}{\partial u^3} \\ \frac{\partial x^2}{\partial u^1} & \frac{\partial x^2}{\partial u^2} & \frac{\partial x^2}{\partial u^3} \\ \frac{\partial x^3}{\partial u^1} & \frac{\partial x^3}{\partial u^2} & \frac{\partial x^3}{\partial u^3} \end{bmatrix}$$

The entries of the product of these matrices then correspond to the entries of the metric tensor as given by Eq. 216.

2.6 Basis Vectors and Their Relation to the Metric and Jacobian

Moreover, the determinant of a matrix is equal to the determinant of its transpose, that is:

$$\det(\mathbf{J}^T) = \det(\mathbf{J}) = J \tag{65}$$

As explained earlier, in orthogonal coordinate systems the covariant and contravariant basis vectors, $\mathbf{E}_i$ and $\mathbf{E}^i$, at any point of the space are in the same direction, and hence the normalization of each one of these basis sets, by dividing each basis vector by its magnitude, produces identical orthonormal basis sets. Consequently, there is no difference between the covariant and contravariant components of tensors with respect to such contravariant and covariant orthonormal basis sets. This, however, is not true in general coordinate systems where each normalized basis set is different in general from the other.

When the covariant basis vectors $\mathbf{E}_i$ are mutually orthogonal at each point of the space, the following consequences will follow:

1. The contravariant basis vectors $\mathbf{E}^i$ are also mutually orthogonal because the corresponding vectors of each basis set are in the same direction due to the fact that the tangent vector to the ith coordinate curve and the gradient vector of the ith coordinate surface at a given point in the space have the same direction. This may also be established more formally by Eq. 49 since the reciprocal of an orthogonal system of vectors should also be orthogonal. Eqs. 50 and 51 may also be used in this argument since the cross products in the numerators are orthogonal to the multiplicands.

2. The covariant and contravariant metric tensors, g_{ij} and g^{ij}, are diagonal with non-vanishing diagonal elements, that is:[12]

$$g_{ij} = 0 \qquad g^{ij} = 0 \qquad (i \neq j) \tag{66}$$

$$g_{ii} \neq 0 \qquad g^{ii} \neq 0 \qquad (\text{no sum on } i) \tag{67}$$

This can be concluded from Eq. 48 since the dot product is zero when the indices are different due to orthogonality. Also, the dot product should be non-zero when the indices are identical because the basis vectors cannot vanish at the regular points of the space since the tangent to the coordinate curve and the gradient to the coordinate surface do exist and they cannot be zero. Moreover, the metric tensor (as we will see in § 4.5) is invertible and hence it does not vanish at any point in the space. This guarantees that none of the diagonal elements can be zero when the metric tensor is diagonal since the determinant, which is formed by the product of its diagonal elements, should not vanish.

3. The diagonal elements of the covariant and contravariant metric tensors are reciprocals, that is:

$$g^{ii} = \frac{1}{g_{ii}} \qquad (\text{no summation}) \tag{68}$$

This is a result from the previous point (i.e. the metric tensor is diagonal with non-vanishing diagonal elements) since the covariant and contravariant metric tensors

---

[12] This also applies to the mixed type metric tensor since it is the identity tensor according to Eq. 49 although this is not restricted to orthogonal coordinate systems.

are inverses of each other. As it is well known from linear algebra, the inverse of a diagonal matrix with non-vanishing diagonal entries is a diagonal matrix obtained by taking the reciprocals of its diagonal elements.

4. The magnitudes of the contravariant and covariant basis vectors are reciprocals, that is:

$$\left|\mathbf{E}^i\right| = \frac{1}{|\mathbf{E}_i|} = \frac{1}{h_i} \tag{69}$$

This result is based on the definition of vector modulus plus Eqs. 48, 68 and 53, that is:

$$\begin{aligned}
\left|\mathbf{E}^i\right| &= \sqrt{\mathbf{E}^i \cdot \mathbf{E}^i} & \text{(Eq. 263)} \\
&= \sqrt{g^{ii}} & \text{(Eq. 48)} \\
&= \frac{1}{\sqrt{g_{ii}}} & \text{(Eq. 68)} \\
&= \frac{1}{\sqrt{\mathbf{E}_i \cdot \mathbf{E}_i}} & \text{(Eq. 48)} \\
&= \frac{1}{|\mathbf{E}_i|} & \text{(Eq. 264)} \\
&= \frac{1}{h_i} & \text{(Eq. 53)}
\end{aligned} \tag{70}$$

with no sum over i.

The consequences that we stated in the above bullet points also follow when the contravariant basis vectors $\mathbf{E}^i$ are mutually orthogonal with minor amendments required in the phrasing of the first point.

2.7 Relationship between Space, Coordinates and Metric

In this section, we try to shed some light on the relationship between space, coordinate system and metric tensor where in this discussion we need the concepts of basis vectors and coordinate curves and coordinate surfaces (see § 2.4 and 2.6).[13] We start from the very abstract concept of space (say nD space) which may be visualized as a sort of vacuum in which mathematical objects can exist. The number n represents the dimensionality of the space which is a quantitative measure of the complexity of the space structure where an nD space requires exactly n mutually independent variables to be fully and uniquely described. So, any point in this nD space can be uniquely located by assigning an ordered set of n mutually independent variables which we call "coordinates" such as $u^1, \cdots, u^n$. These coordinates vary continuously throughout the space where each coordinate ranges over a given real interval which may be finite or infinite. All the points of the space are then uniquely identified by varying these variables over their predefined ranges. The way

---

[13] The revelation in this section is highly informal and non-rigorous with a pedagogical objective of providing a qualitative appreciation of the relation between space, coordinate system and metric tensor which may be useful for some novice readers.

2.7 Relationship between Space, Coordinates and Metric

by which the points of space are identified by the ordered sets of n coordinates may be labeled generically as a "coordinate system" for the space.

The condition of mutual independence of the coordinate variables is important to make the distinction between the points unique and universal and to avoid any ambiguity. Mutual independence requires that an nD space is identified by a coordinate system with exactly n variable coordinates. As well as the condition of mutual independence, there is another important condition for the coordinate system to be acceptable that is it should be universal in its ability to identify all the points of the space unambiguously, and hence the coordinate system should be thorough in its functionality with possible exclusion of a finite number of isolated points in the space. The "coordinate system" may then be defined more rigorously as a bijective correspondence between the set of n-tuple numbers and the points of the space.

For the coordinate system to be of any use, the way by which the coordinates are assigned to the individual points should be clearly defined. So, each particular way of defining the aforementioned bijective correspondence represents a specific coordinate system associated with the space. In fact, , there are infinitely many ways of choosing the coordinate variables and how they vary in the space, and hence there are infinitely many coordinate systems that can be chosen to fulfill the above objectives. However, the conditions of mutual independence of coordinates and bijectivity of mapping impose certain restrictions on the selection of coordinates and how they can vary to achieve the intended functionalities.

As a result of the above conditions and restrictions on the coordinate variables and how they vary to ensure their mutual independence and universal validity at each point in the space, there are only certain allowed ways for the choice of the coordinate curves and coordinate surfaces. So, on excluding certain types of coordinate systems which do not meet these requirements, and hence excluding certain types of how the coordinate curves and surfaces are defined at the individual points of the space, we are left with only certain acceptable possibilities for identifying how the coordinates vary and hence how the coordinate curves and surfaces are identified.

Now, since the covariant basis vectors are defined as the tangents to the coordinate curves and the contravariant basis vectors are defined as the gradients to the coordinate surfaces, then to each one of those acceptable possibilities there is only one possibility for the identification of the covariant and contravariant sets of basis vectors. As soon as the two sets of basis vectors of the selected coordinate system are identified unambiguously, the metric tensor in its covariant and contravariant forms is uniquely identified throughout the space through its relation to the basis vectors, i.e. Eq. 48, and hence the vagueness about the nature of the space, its geometry and how it should be described is removed thanks to the existence of a well defined metric tensor.

In brief, we start by defining a coordinate system which leads to the definition of coordinate curves and coordinate surfaces (and hence covariant and contravariant basis vector sets) and this in its turn will lead to the definition of the metric tensor. In fact, this approach may be described as procedural since it is based on the practical procedure of defining these concepts. A more fundamental approach may start from the metric of the space as an essential property of the space that characterizes its nature from which the basis vectors and the coordinate system are to be derived. It should be remarked that the coordinate system and the space metric are independent entities despite the close link between the two. The metric can be regarded as a built-in attribute that characterizes the space but it can be demonstrated in different forms depending on the employed coordinate system.

In this context, it is important to distinguish between the intrinsic and extrinsic properties of an nD space where the intrinsic properties refer to those properties of the space which are seen from inside the space by an inhabitant of the space who can perceive only the n dimensions of the space, while the extrinsic properties refer to those properties of the space which are seen from outside the space by an inhabitant of an mD space $(m > n)$ that encloses the nD space where this mD inhabitant can perceive the n dimensions of the space plus the extra $(m - n)$ dimensions of the mD space. While the intrinsic properties of the nD space are described and quantified by the metric of the nD space, the extrinsic properties of the space are described and quantified by the metric of the embedding mD space since the nD space is a subspace of the mD space and hence it is subject to the mD metric. An example of nD and mD spaces is a 2D surface, such as a plane or a sphere, which is embedded in a 3D Euclidean space.

2.8 Exercises and Revision

2.1 Give brief definitions to the following terms: Riemannian space, coordinate system and metric tensor.

2.2 Discuss the main functions of the metric tensor in a given space. How many types the metric tensor can have?

2.3 What is the meaning of "flat" and "curved" space? Give mathematical conditions for the space to be flat in terms of the length of an infinitesimal element of arc and in terms of the metric tensor. Why these conditions should be global for the space to be flat?

2.4 Give common examples of flat and curved spaces of different dimensions justifying in each case why the space is flat or curved.

2.5 Explain why all 1D spaces are Euclidean.

2.6 Give examples of spaces with constant curvature and spaces with variable curvature.

2.7 State Schur theorem outlining its significance.

2.8 What is the condition for a space to be intrinsically flat and extrinsically flat?

2.9 What is the common method of investigating the Riemannian geometry of a curved manifold?

2.10 Give brief definitions to coordinate curves and coordinate surfaces outlining their relations to the basis vector sets. How many independent coordinate curves and coordinate surfaces we have at each point of a 3D space with a valid coordinate system?

2.11 Why a coordinate system is needed in tensor formulations?

2.12 List the main types of coordinate system outlining their relations to each other.

2.13 "The coordinates of a system can have the same physical dimension or different phys-

2.8 Exercises and Revision

ical dimensions". Give an example for each.

2.14 Prove that spherical coordinate systems are orthogonal.

2.15 What is the difference between rectilinear and curvilinear coordinate systems?

2.16 Give examples of common curvilinear coordinate systems explaining why they are curvilinear.

2.17 Give an example of a commonly used curvilinear coordinate system with some of its coordinate curves being straight lines.

2.18 Define briefly the terms "orthogonal" and "homogeneous" coordinate system.

2.19 Give examples of rectilinear and curvilinear orthogonal coordinate systems.

2.20 What is the condition of a coordinate system to be orthogonal in terms of the form of its metric tensor? Explain why this is so.

2.21 What is the mathematical condition for a coordinate system to be homogeneous?

2.22 How can we homogenize a non-homogeneous coordinate system of a flat space?

2.23 Give examples of homogeneous and non-homogeneous coordinate systems.

2.24 Give an example of a non-homogeneous coordinate system that can be homogenized.

2.25 Describe briefly the transformation of spaces and coordinate systems stating relevant mathematical relations.

2.26 What "injective transformation" means? Is it necessary that such a transformation has an inverse?

2.27 Write the Jacobian matrix $\mathbf{J}$ of a transformation between two nD spaces whose coordinates are labeled as u^i and $\bar{u}^i$ where $i = 1, \cdots, n$.

2.28 State the pattern of the row and column indices of the Jacobian matrix in relation to the indices of the coordinates of the two spaces.

2.29 What is the difference between the Jacobian matrix and the Jacobian and what is the relation between them?

2.30 What is the relation between the Jacobian of a given transformation and the Jacobian of its inverse? Write a mathematical formula representing this relation.

2.31 Is the labeling of two coordinate systems (e.g. barred and unbarred) involved in a transformation relation essential or arbitrary? Hence, discuss if the labeling of the coordinates in the Jacobian matrix can be interchanged.

2.32 Using the transformation equations between the Cartesian and cylindrical coordinate systems, find the Jacobian matrix of the transformation between these systems, i.e. Cartesian to cylindrical and cylindrical to Cartesian.

2.33 Repeat question 2.32 for the spherical, instead of cylindrical, system to find the Jacobian this time.

2.34 Give a simple definition of admissible coordinate transformation.

2.35 What is the meaning of the C^n continuity condition?

2.36 What "invariant" object or property means? Give some illustrating examples.

2.37 What is the meaning of "composition of transformations"? State a mathematical relation representing such a composition.

2.38 What is the Jacobian of a composite transformation in terms of the Jacobians of the simple transformations that make the composite transformation? Write a mathematical relation that links all these Jacobians.

2.8 Exercises and Revision

2.39 "The collection of all admissible coordinate transformations with non-vanishing Jacobian form a group". What this means? State your answer in mathematical and descriptive forms.

2.40 Is the transformation of coordinates a commutative operation? Justify your answer by an example.

2.41 A transformation T_3 with a Jacobian J_3 is a composite transformation, i.e. $T_3 = T_2 T_1$ where the transformations T_1 and T_2 have Jacobians J_1 and J_2. What is the relation between J_1, J_2 and J_3?

2.42 Two transformations, R_1 and R_2, are related by: $R_1 R_2 = I$ where I is the identity transformation. What is the relation between the Jacobians of R_1 and R_2? What we should call these transformations?

2.43 Discuss the transformation of one set of basis vectors of a given coordinate system to another set of opposite variance type of that system and the relation of this to the metric tensor.

2.44 Discuss the transformation of one set of basis vectors of a given coordinate system to another set of the same variance type of another coordinate system.

2.45 Discuss and compare the results of question 2.43 and question 2.44. Also, compare the mathematical formulation that should apply in each case.

2.46 Define proper and improper coordinate transformations.

2.47 What is the difference between positive and negative orthogonal transformations?

2.48 Give detailed definitions of coordinate curves and coordinate surfaces of 3D spaces discussing the relation between them.

2.49 For each one of the following coordinate systems, what is the shape of the coordinate curves and coordinate surfaces: Cartesian, cylindrical and spherical?

2.50 Make a simple plot representing the ϕ coordinate curve with the ρ and z coordinate surfaces of a cylindrical coordinate system.

2.51 Make a simple plot representing the r coordinate curve with the θ and ϕ coordinate surfaces of a spherical coordinate system.

2.52 Define "scale factors" of a coordinate system and outline their significance.

2.53 Give the scale factors of the following coordinate systems: orthonormal Cartesian, cylindrical and spherical.

2.54 Define, mathematically and in words, the covariant and contravariant basis vector sets explaining any symbols involved in these definitions.

2.55 What is the relation of the covariant and contravariant basis vector sets with the coordinate curves and coordinate surfaces of a given coordinate system? Make a simple sketch representing this relation for a general curvilinear coordinate system in a 3D space.

2.56 The covariant and contravariant components of vectors can be transformed one to the other. How? State your answer in a mathematical form.

2.57 What is the significance of the following relations?

$$\mathbf{E}_i \cdot \mathbf{E}^j = \delta_i^j \qquad \mathbf{E}^i \cdot \mathbf{E}_j = \delta^i{}_j$$

2.58 Write down the mathematical relations that correlate the basis vectors to the compo-

2.8 Exercises and Revision

nents of the metric tensor in their covariant and contravariant forms.

2.59 Using Eq. 50, show that if $\mathbf{E}^1, \mathbf{E}^2, \mathbf{E}^3$ is a right handed orthonormal system then $\mathbf{E}_i = \mathbf{E}^i$. Repeat the question using this time Eq. 51 where $\mathbf{E}_1, \mathbf{E}_2, \mathbf{E}_3$ form a right handed orthonormal system. Hence, conclude that when the covariant or contravariant basis vector set is orthonormal then the covariant and contravariant components of a given tensor are identical.

2.60 State the mathematical relations between the original and transformed (i.e. unbarred and barred) basis vector sets in their covariant and contravariant forms under admissible coordinate transformations.

2.61 Correct, if necessary, the following equations explaining all the symbols involved:

$$\mathbf{E}_1 \cdot (\mathbf{E}_2 \times \mathbf{E}_3) = \frac{1}{\sqrt{g}} \qquad \mathbf{E}^1 \cdot (\mathbf{E}^2 \times \mathbf{E}^3) = \sqrt{g}$$

2.62 Obtain the relation: $g = J^2$ from the relation: $\mathbf{J}^T \mathbf{J} = [g_{ij}]$ giving full explanation of each step.

2.63 State three consequences of having mutually orthogonal contravariant basis vectors at each point in the space justifying these consequences.

2.64 Discuss the relationship between the concepts of space, coordinate system and metric tensor.

Chapter 3
Tensors

The subject of this chapter is tensors in general. The chapter is divided into three sections where we discuss in these sections: tensor types, tensor operations, and tensor representations.

3.1 Tensor Types

In the following subsections, we introduce a number of tensor types and categories and highlight their main characteristics and differences. These types and categories are not mutually exclusive and hence they overlap in general. Moreover, they may not be exhaustive in their classes as some tensors may not instantiate any one of a complementary set of types such as being symmetric or anti-symmetric.

3.1.1 Covariant and Contravariant Tensors

These are the main types of tensor with regard to the rules of their transformation between different coordinate bases. Covariant tensors are notated with subscript indices (e.g. A_i) while contravariant tensors are notated with superscript indices (e.g. A^{ij}). A covariant tensor is transformed according to the following rule:

$$\bar{A}_i = \frac{\partial u^j}{\partial \bar{u}^i} A_j \tag{71}$$

while a contravariant tensor is transformed according to the following rule:

$$\bar{A}^i = \frac{\partial \bar{u}^i}{\partial u^j} A^j \tag{72}$$

The barred and unbarred symbols in these equations represent the same mathematical object (tensor or coordinate) in the transformed and original coordinate systems respectively. An example of covariant tensors is the gradient of a scalar field while an example of contravariant tensors is the displacement vector.

The above transformation rules, as demonstrated by Eqs. 71 and 72, which correspond to rank-1 tensors can be easily extended to tensors of any rank following the above pattern. Hence, a covariant tensor $\mathbf{A}$ of rank m is transformed according to the following rule:

$$\bar{A}_{ij\cdots m} = \frac{\partial u^p}{\partial \bar{u}^i} \frac{\partial u^q}{\partial \bar{u}^j} \cdots \frac{\partial u^r}{\partial \bar{u}^m} A_{pq\cdots r} \tag{73}$$

Similarly, a contravariant tensor $\mathbf{B}$ of rank n is transformed according to the following rule:

$$\bar{B}^{ij\cdots n} = \frac{\partial \bar{u}^i}{\partial u^p} \frac{\partial \bar{u}^j}{\partial u^q} \cdots \frac{\partial \bar{u}^n}{\partial u^r} B^{pq\cdots r} \tag{74}$$

3.1.1 Covariant and Contravariant Tensors

Some tensors of rank > 1 have mixed variance type, i.e. they are covariant in some indices and contravariant in others. In this case the covariant variables are indexed with subscripts while the contravariant variables are indexed with superscripts, e.g. A_i^j which is covariant in i and contravariant in j. A mixed type tensor transforms covariantly in its covariant indices and contravariantly in its contravariant indices, e.g.

$$\bar{A}^l{}_m{}^n = \frac{\partial \bar{u}^l}{\partial u^i} \frac{\partial u^j}{\partial \bar{u}^m} \frac{\partial \bar{u}^n}{\partial u^k} A^i{}_j{}^k \tag{75}$$

In brief, tensors of rank-n $(n > 0)$ whose all free indices are subscripts are covariant, and tensors of rank-n $(n > 0)$ whose all free indices are superscripts are contravariant, while tensors of rank-n $(n > 1)$ whose some of free indices are subscripts while other free indices are superscripts are mixed. Hence, A_i and B_{ij} are covariant and transform according to Eq. 73, A^i and B^{ijk} are contravariant and transform according to Eq. 74, and A_i^j and B_i^{jk} are mixed and transform according to Eq. 75. As indicated before, for orthonormal Cartesian coordinate systems there is no difference between the covariant and contravariant representations, and hence the indices can be upper or lower although it is common to use lower indices in this case.

The practical rules for writing the transformation equations, as seen in the above equations, can be summarized in the following points where we take Eq. 75 (which is sufficiently general) as an example to illustrate the steps taken in writing the transformation equations:

1. We start by writing the symbol of the transformed tensor on the left hand side of the transformation equation and the symbol of the original tensor on the right hand side, that is:

$$\bar{A} = A \tag{76}$$

2. We index the transformed tensor according to its intended indicial structure observing the order of the indices. We similarly index the original tensor with its original indices, that is:

$$\bar{A}^l{}_m{}^n = A^i{}_j{}^k \tag{77}$$

3. We insert a number of partial differential operators on the right hand side equal to the number of free indices, that is:

$$\bar{A}^l{}_m{}^n = \frac{\partial u}{\partial u} \frac{\partial u}{\partial u} \frac{\partial u}{\partial u} A^i{}_j{}^k \tag{78}$$

4. We index the coordinates of the transformed tensor in the numerator or denominator in these operators according to the order of the indices in the tensor where these indices are in the same position (upper or lower) as their position in the tensor, that is:

$$\bar{A}^l{}_m{}^n = \frac{\partial u^l}{\partial u} \frac{\partial u}{\partial u^m} \frac{\partial u^n}{\partial u} A^i{}_j{}^k \tag{79}$$

5. Because the transformed tensor is barred then its coordinates should also be barred, that is:

$$\bar{A}^l{}_m{}^n = \frac{\partial \bar{u}^l}{\partial u} \frac{\partial u}{\partial \bar{u}^m} \frac{\partial \bar{u}^n}{\partial u} A^i{}_j{}^k \tag{80}$$

3.1.1 Covariant and Contravariant Tensors

6. We then index the coordinates of the original tensor in the numerator or denominator in these operators according to the order of the indices in the tensor where these indices are in the opposite position (upper or lower) to their position in the tensor, that is:

$$\bar{A}^l{}_m{}^n = \frac{\partial \bar{u}^l}{\partial u^i} \frac{\partial u^j}{\partial \bar{u}^m} \frac{\partial \bar{u}^n}{\partial u^k} A^i{}_j{}^k \tag{81}$$

We note that an upper coordinate index in the denominator of the partial differential operators is equivalent to a lower tensor index. We also note that each pair of the transformed indices are on the two sides and hence they should be in the same position (upper or lower) according to the rules of free indices (see § 1.2), while the pairs of indices of the original tensor are in the same side and hence they should be in opposite positions according to the rules of dummy indices.

The covariant and contravariant types of a tensor are linked through the metric tensor. As will be detailed later (refer to § 4.5), the contravariant metric tensor is used for raising covariant indices of covariant and mixed tensors, e.g.

$$A^i = g^{ik} A_k \qquad\qquad A^{ij} = g^{ik} A_k{}^j \tag{82}$$

where the contravariant metric tensor g^{ik} is used to raise the covariant index k to become the contravariant index i. Similarly, the covariant metric tensor is used for lowering contravariant indices of contravariant and mixed tensors, e.g.

$$A_i = g_{ik} A^k \qquad\qquad A_{ij} = g_{ik} A^k{}_j \tag{83}$$

where the covariant metric tensor g_{ik} is used to lower the contravariant index k to become the covariant index i. Hence, the metric tensor is used to change the variance type of tensor indices and for that reason it is labeled as index raising or index lowering or index shifting operator. We note that in the raising and lowering operations, the metric tensor acts as an index replacement operator, as well as a shifting operator, by changing the label of the shifted index, as seen in the above examples where the shifted index k is replaced with i which is the other index of the metric tensor.

Because it is possible to shift the index position of a tensor by using the covariant and contravariant types of the metric tensor as an index shifting operator, a given tensor can be cast into a covariant or a contravariant form, as well as a mixed form in the case of tensors of rank > 1. However, it should be emphasized that the order of the indices must be respected in this process, because two tensors with the same indicial structure but with different indicial order are not equal in general, as stated before. For example:

$$A^i{}_j = g_{jk} A^{ik} \neq g_{jk} A^{ki} = A_j{}^i \tag{84}$$

Dots may be inserted to remove any ambiguity about the order of the indices and hence $A^i{}_{\cdot j}$ or $A^i{}_j$ means i first and j second while $A_j{}^{\cdot i}$ or $A_j{}^i$ means j first and i second.

A tensor of m free contravariant indices and n free covariant indices may be called type (m,n) tensor. When one or both variance types are absent, zero is used to refer to the absent variance type in this notation. Accordingly, A^k_{ij} is a type $(1,2)$ tensor, B^{ik} is a

3.1.1 Covariant and Contravariant Tensors

type $(2,0)$ tensor, C_m is a type $(0,1)$ tensor, and D^{ts}_{pqr} is a type $(2,3)$ tensor. As we will see in § 3.1.3, the type may also include the weight as a third entry and hence the type in this sense is identified by the symbol (m,n,w) where m and n refer to the number of contravariant and covariant indices respectively while w refers to the weight of the tensor. It is obvious that the rank of the tensor can be obtained from this notation by adding the first two entries and hence a tensor of type (m,n) or (m,n,w) is of rank $m+n$.

As seen earlier, the vectors providing the basis set for a coordinate system are of covariant type when they are tangent to the coordinate curves, which represent the curves along which exactly one coordinate varies while the other coordinates are held constant, and they are of contravariant type when they are perpendicular to the coordinate surfaces which are the surfaces over which all coordinates vary except one which is held constant. Formally, the covariant and contravariant basis vectors are given respectively by:

$$\mathbf{E}_i = \frac{\partial \mathbf{r}}{\partial u^i} \qquad \mathbf{E}^i = \nabla u^i \qquad (85)$$

where $\mathbf{r} = x_i \mathbf{e}_i$ is the position vector in Cartesian coordinates and u^i is a general coordinate. As indicated earlier, a superscript in the denominator of partial derivatives is equivalent to a subscript in the numerator and hence the above equation is consistent with the rules of tensor indices which were outlined in § 1.2.

We also remark that in general the basis vectors, whether covariant or contravariant, are not necessarily of unit length and/or mutually orthogonal although they may be so. In fact, there are standard mathematical procedures to orthonormalize the basis set if it is not and if orthonormalization is needed. For example, the two covariant basis vectors, $\mathbf{E}_1$ and $\mathbf{E}_2$, of a 2D space (i.e. surface) can be orthonormalized as follows:

$$\underline{\mathbf{E}}_1 = \frac{\mathbf{E}_1}{|\mathbf{E}_1|} = \frac{\mathbf{E}_1}{\sqrt{g_{11}}} \qquad \underline{\mathbf{E}}_2 = \frac{g_{11}\mathbf{E}_2 - g_{12}\mathbf{E}_1}{\sqrt{g_{11}g}} \qquad (86)$$

where g is the determinant of the surface covariant metric tensor, the indexed g are the coefficients of this tensor, and the underlined vectors are orthonormal basis vectors, that is:

$$\underline{\mathbf{E}}_1 \cdot \underline{\mathbf{E}}_1 = 1 \qquad \underline{\mathbf{E}}_2 \cdot \underline{\mathbf{E}}_2 = 1 \qquad \underline{\mathbf{E}}_1 \cdot \underline{\mathbf{E}}_2 = 0 \qquad (87)$$

This can be verified by conducting the dot products of the last equation using the vectors defined in Eq. 86.

As seen earlier, the two sets of covariant and contravariant basis vectors are reciprocal systems and hence they satisfy the following reciprocity relations:

$$\mathbf{E}_i \cdot \mathbf{E}^j = \delta_i^j \qquad \mathbf{E}^i \cdot \mathbf{E}_j = \delta^i_{\ j} \qquad (88)$$

where the indexed δ is the Kronecker delta tensor (refer to § 4.1) which can be represented by the unity matrix. The reciprocity of these two sets of basis vectors is illustrated schematically in Figure 11 for the case of a 2D space.

3.1.1 Covariant and Contravariant Tensors

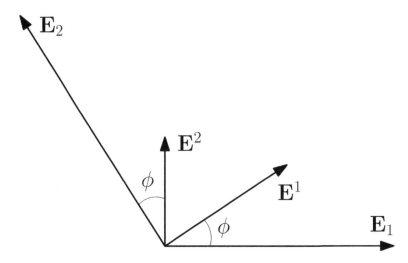

Figure 11: The reciprocity relation between the covariant and contravariant basis vectors in a 2D space where $\mathbf{E}_1 \perp \mathbf{E}^2$, $\mathbf{E}^1 \perp \mathbf{E}_2$, and $|\mathbf{E}_1||\mathbf{E}^1|\cos\phi = |\mathbf{E}_2||\mathbf{E}^2|\cos\phi = 1$.

A vector can be represented either by covariant components with contravariant basis vectors or by contravariant components with covariant basis vectors. For example, a vector $\mathbf{A}$ can be expressed as:

$$\mathbf{A} = A_i \mathbf{E}^i \qquad \text{or} \qquad \mathbf{A} = A^i \mathbf{E}_i \tag{89}$$

where $\mathbf{E}^i$ and $\mathbf{E}_i$ are the contravariant and covariant basis vectors respectively. This is illustrated graphically in Figure 12 for a vector $\mathbf{A}$ in a 2D space. The use of the covariant or contravariant form of the vector representation is a matter of choice and convenience since these two representations are equivalent as they represent and correctly describe the same object.

More generally, a tensor of any rank (≥ 1) can be represented covariantly using contravariant basis tensors of that rank, or contravariantly using covariant basis tensors, or in a mixed form using a mixed basis of opposite type. For example, a rank-2 tensor $\mathbf{A}$ can be written as:

$$\mathbf{A} = A_{ij}\mathbf{E}^i\mathbf{E}^j = A^{ij}\mathbf{E}_i\mathbf{E}_j = A_i{}^j\mathbf{E}^i\mathbf{E}_j = A^i{}_j\mathbf{E}_i\mathbf{E}^j \tag{90}$$

where $\mathbf{E}^i\mathbf{E}^j$, $\mathbf{E}_i\mathbf{E}_j$, $\mathbf{E}^i\mathbf{E}_j$ and $\mathbf{E}_i\mathbf{E}^j$ are dyadic products of the basis vectors of the presumed system (refer to § 3.2.3). We remark that dyadic products represent a combination of two vectors and hence they represent two directions in a certain order. Figure 13 is a graphic illustration of the nine dyadic products of the three unit basis vectors in a 3D space with a rectangular Cartesian coordinate system.

Similarly, a rank-3 tensor $\mathbf{B}$ can be written as:

$$\mathbf{B} = B_{ijk}\mathbf{E}^i\mathbf{E}^j\mathbf{E}^k = B^{ijk}\mathbf{E}_i\mathbf{E}_j\mathbf{E}_k = B_i{}^{jk}\mathbf{E}^i\mathbf{E}_j\mathbf{E}_k = B^i{}_j{}^k\mathbf{E}_i\mathbf{E}^j\mathbf{E}_k = \cdots \text{etc.} \tag{91}$$

where $\mathbf{E}^i\mathbf{E}^j\mathbf{E}^k\ldots$ etc. are triads. More generally, a rank-n tensor $\mathbf{C}$ can be written as:

$$\mathbf{C} = C_{ij\cdots n}\mathbf{E}^i\mathbf{E}^j\cdots\mathbf{E}^n = C^{ij\cdots n}\mathbf{E}_i\mathbf{E}_j\cdots\mathbf{E}_n = C_i{}^{j\cdots n}\mathbf{E}^i\mathbf{E}_j\cdots\mathbf{E}_n = \cdots \text{etc.} \tag{92}$$

3.1.2 True and Pseudo Tensors 55

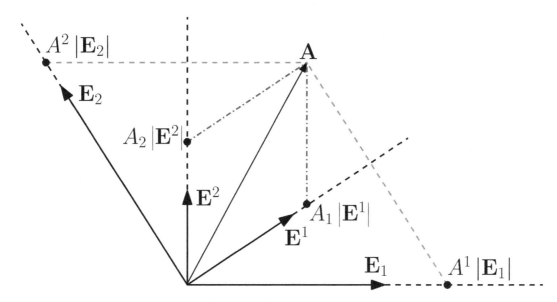

Figure 12: The representation of a vector **A** in covariant and contravariant basis vector sets in a 2D space where the components shown at the four points are with reference to unit vectors in the given directions, e.g. $A^1 |\mathbf{E}_1|$ is a component with reference to a unit vector in the direction of $\mathbf{E}_1$.

where $\mathbf{E}^i \mathbf{E}^j \cdots \mathbf{E}^n \ldots$ etc. are n-polyads. We note that the order of the indices in the tensor bases (i.e. dyads, triads and n-polyads) should be the same as the order of the indices in the tensor components, as seen in the above examples.

Finally, it should be remarked that the two sets of basis vectors (i.e. covariant and contravariant), like the components of the tensors themselves, are identical for orthonormal Cartesian systems. This can be explained by the fact that since the coordinate surfaces are mutually perpendicular planes then the coordinate curves are mutually perpendicular straight lines and hence the tangent vectors to the x^i coordinate curve is the same as the perpendicular vector to the x^i coordinate surface noting that the coordinates are scaled uniformly in all directions.

3.1.2 True and Pseudo Tensors

These are also called polar and axial tensors respectively although it is more common to use these terms for vectors. Pseudo tensors may also be called tensor densities, however the terminology in this part, like many other parts, is not universal. True tensors are proper or ordinary tensors and hence they are invariant under coordinate transformations, while pseudo tensors are not proper tensors since they do not transform invariantly as they acquire a minus sign under improper orthogonal transformations which involve inversion of coordinate axes through the origin of coordinates with a change of system handedness. Figure 14 demonstrates the behavior of a true vector **v** and a pseudo vector **p** where the former keeps its direction in the space following a reflection of the coordinate system through the origin of coordinates while the latter reverses its direction following

3.1.2 True and Pseudo Tensors

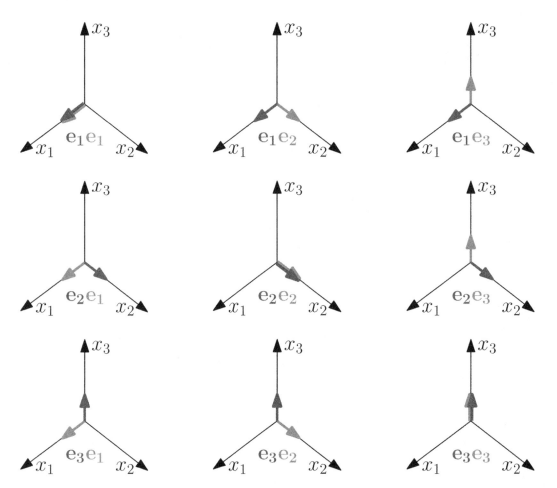

Figure 13: The nine unit dyads associated with the double directions of rank-2 tensors in a 3D space with a rectangular Cartesian coordinate system. The vectors e_i and e_j ($i, j = 1, 2, 3$) are unit vectors in the directions of coordinate axes where the first indexed **e** (blue) represents the first vector of the dyad while the second indexed **e** (red) represents the second vector of the dyad. In these nine frames, the first vector is fixed along each row while the second vector is fixed along each column.

this operation.

Because true and pseudo tensors have different mathematical properties and represent different types of physical entities, the terms of consistent tensor expressions and equations should be uniform in their true and pseudo type, i.e. all terms should be true or all terms should be pseudo. The direct product (refer to § 3.2.3) of true tensors is a true tensor. The direct product of even number of pseudo tensors is a true tensor, while the direct product of odd number of pseudo tensors is a pseudo tensor. The direct product of a mix of true and pseudo tensors is a true or pseudo tensor depending on the number of pseudo tensors involved in the product as being even or odd respectively.

Similar rules to those of the direct product apply to the cross product, including the curl operation, involving tensors (which are usually of rank-1) with the addition of a pseudo

3.1.3 Absolute and Relative Tensors

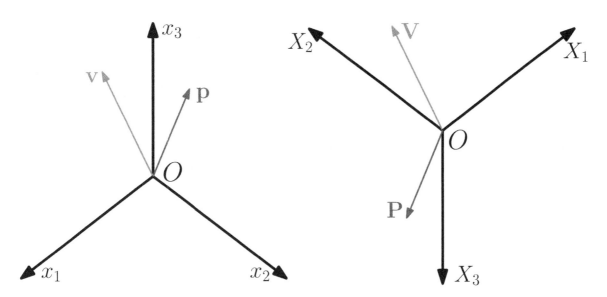

Figure 14: The behavior of a true vector (**v** and **V**) and a pseudo vector (**p** and **P**) on reflecting the coordinate system in the origin of coordinates. The lower case symbols stand for the objects in the original system while the upper case symbols stand for the same objects in the reflected system.

factor for each cross product operation. This factor is contributed by the permutation tensor which is implicit in the definition of the cross product. As we will see in § 4.2, the permutation tensor is a pseudo tensor.

In summary, what determines the tensor type (true or pseudo) of the tensor terms involving direct and cross products is the parity of the multiplicative factors of pseudo type plus the number of cross product operations involved since each cross product operation contributes a permutation tensor. We note that inner product (see § 3.2.5) is the result of a direct product (see § 3.2.3) operation followed by a contraction (see § 3.2.4) and hence it is like a direct product in this context.

3.1.3 Absolute and Relative Tensors

Considering an arbitrary transformation from a general coordinate system to another, a tensor of weight w is defined by the following general tensor transformation relation:

$$\bar{A}^{ij\ldots k}_{lm\ldots n} = \left|\frac{\partial u}{\partial \bar{u}}\right|^w \frac{\partial \bar{u}^i}{\partial u^a}\frac{\partial \bar{u}^j}{\partial u^b}\cdots\frac{\partial \bar{u}^k}{\partial u^c}\frac{\partial u^d}{\partial \bar{u}^l}\frac{\partial u^e}{\partial \bar{u}^m}\cdots\frac{\partial u^f}{\partial \bar{u}^n}A^{ab\ldots c}_{de\ldots f} \qquad (93)$$

where $\left|\frac{\partial u}{\partial \bar{u}}\right|$ symbolizes the Jacobian of the transformation between the two systems (see § 2.3). When $w = 0$ the tensor is described as an absolute or true tensor, and when $w \neq 0$ the tensor is described as a relative tensor. When $w = -1$ the tensor may be described as a pseudo tensor, while when $w = 1$ the tensor may be described as a tensor density. We note that some of these labels are used differently by different authors as the terminology of tensor calculus is not universally approved and hence the conventions of each author should be checked. Also, there is an obvious overlap between this classification (i.e. absolute and relative) and the previous classification (i.e. true and pseudo) at least according to some conventions. As indicated earlier (see § 3.1.1), a tensor of m contravariant indices and n covariant indices may be described as a tensor of type (m, n). This may be extended to include the weight w as a third entry and hence the type of the tensor is identified by (m, n, w).

Relative tensors can be added and subtracted (see § 3.2.1) if they have the same indicial structure and have the same weight; the result is a tensor of the same indicial structure and weight. Also, relative tensors can be equated if they have the same indicial structure and weight. In brief, the terms of tensor expressions and equalities should have the same weight w, whether w is equal to zero or not. Multiplication of relative tensors produces a relative tensor whose weight is the sum of the weights of the original tensors. Hence, if the weights are added up to a non-zero value the result is a relative tensor of that weight; otherwise it is an absolute tensor.

We remark that the statements of the previous paragraph can be generalized by including $w = 0$ which corresponds to absolute tensors and hence "relative" in those statements is more general than being opposite to "absolute". Accordingly, and from the perspective of relative tensors (i.e. assuming that other qualifications such as matching in indicial structure, are met), two absolute tensors can be added or subtracted or equated but an absolute and a relative tensor (i.e. with $w \ne 0$) cannot since they are "relative" tensors with different weights.

3.1.4 Isotropic and Anisotropic Tensors

Isotropic tensors are characterized by the property that the values of their components are invariant under coordinate transformation by proper rotation of axes. In contrast, the values of the components of anisotropic tensors are dependent on the orientation of the coordinate axes. Notable examples of isotropic tensors are scalars (rank-0), the vector **0** (rank-1), Kronecker delta (rank-2) and Levi-Civita tensor (rank ≥ 2). Many tensors describing physical properties of materials, such as stress and magnetic susceptibility, are anisotropic.

Direct and inner products (see § 3.2.3 and 3.2.5) of isotropic tensors are isotropic tensors. The zero tensor of any rank and any dimension is isotropic; therefore if the components of a tensor vanish in a particular coordinate system they will vanish in all properly and improperly rotated coordinate systems.[14] Consequently, if the components of two tensors are identical in a particular coordinate system they are identical in all transformed coordinate systems since the tensor of their difference is a zero tensor and hence it is invariant. This means that tensor equalities and identities are invariant under coordinate transformations, which is one of the main motivations for the use of tensors in mathematics and science. As indicated, all rank-0 tensors (scalars) are isotropic. Also, the zero vector, **0**, of any dimension is isotropic; in fact it is the only rank-1 isotropic tensor.

---

[14] For improper rotation, this is more general than being isotropic.

3.1.5 Symmetric and Anti-symmetric Tensors

These types of tensor apply to high ranks only (rank ≥ 2) since symmetry and anti-symmetry of tensors require in their definition two free indices at least, and hence a scalar with no index and a vector with a single index do not qualify to be symmetric or anti-symmetric. Moreover, these types are not exhaustive, even for tensors of rank ≥ 2, as there are high-rank tensors which are neither symmetric nor anti-symmetric, and hence we may call them asymmetric tensors although this terminology may include anti-symmetric tensors as well.

A rank-2 tensor A_{ij} is symmetric in its components *iff* for all i and j the following condition is satisfied:

$$A_{ji} = A_{ij} \tag{94}$$

and anti-symmetric or skew-symmetric *iff* for all i and j the following condition is satisfied:

$$A_{ji} = -A_{ij} \tag{95}$$

Similar conditions apply to contravariant type tensors, i.e. $A^{ji} = A^{ij}$ for symmetric and $A^{ji} = -A^{ij}$ for anti-symmetric. This also applies to the following definitions and identities which are largely presented in covariant forms.

More generally, a rank-n tensor $A_{i_1 \ldots i_n}$ is symmetric in its two indices i_j and i_l *iff* the following condition applies identically:

$$A_{i_1 \ldots i_l \ldots i_j \ldots i_n} = A_{i_1 \ldots i_j \ldots i_l \ldots i_n} \tag{96}$$

and anti-symmetric in its two indices i_j and i_l *iff* the following condition applies identically:

$$A_{i_1 \ldots i_l \ldots i_j \ldots i_n} = -A_{i_1 \ldots i_j \ldots i_l \ldots i_n} \tag{97}$$

Any rank-2 tensor A_{ij} can be synthesized from (or decomposed into) a symmetric part $A_{(ij)}$, which is marked with round brackets enclosing the indices, and an anti-symmetric part $A_{[ij]}$, which is marked with square brackets, where the following relations apply:

$$A_{ij} = A_{(ij)} + A_{[ij]} \tag{98}$$

$$A_{(ij)} = \frac{1}{2}\left(A_{ij} + A_{ji}\right) \tag{99}$$

$$A_{[ij]} = \frac{1}{2}\left(A_{ij} - A_{ji}\right) \tag{100}$$

The first relation can be verified by substituting the second and third relations into the first, that is:

$$A_{ij} = A_{(ij)} + A_{[ij]} = \frac{1}{2}\left(A_{ij} + A_{ji}\right) + \frac{1}{2}\left(A_{ij} - A_{ji}\right) = \frac{1}{2}\left(2A_{ij}\right) = A_{ij} \tag{101}$$

which is an identity, while the second and third relations can be verified by shifting the indices, that is:

$$A_{(ji)} = \frac{1}{2}\left(A_{ji} + A_{ij}\right) = \frac{1}{2}\left(A_{ij} + A_{ji}\right) = A_{(ij)} \tag{102}$$

3.1.5 Symmetric and Anti-symmetric Tensors

$$A_{[ji]} = \frac{1}{2}(A_{ji} - A_{ij}) = -\frac{1}{2}(A_{ij} - A_{ji}) = -A_{[ij]} \tag{103}$$

and hence $A_{(ij)}$ is symmetric and $A_{[ij]}$ is anti-symmetric.

Similarly, a rank-3 tensor A_{ijk} can be symmetrized by the following relation:

$$A_{(ijk)} = \frac{1}{3!}(A_{ijk} + A_{kij} + A_{jki} + A_{ikj} + A_{jik} + A_{kji}) \tag{104}$$

and anti-symmetrized by the following relation:

$$A_{[ijk]} = \frac{1}{3!}(A_{ijk} + A_{kij} + A_{jki} - A_{ikj} - A_{jik} - A_{kji}) \tag{105}$$

where the first three terms in these equations are the even permutations of the indices ijk and the last three terms are the odd permutations of these indices. These relations can also be verified by exchanging the indices. For example:

$$A_{(jik)} = \frac{1}{3!}(A_{jik} + A_{kji} + A_{ikj} + A_{jki} + A_{ijk} + A_{kij}) = A_{(ijk)} \tag{106}$$

and hence it is symmetric in i and j. Similarly:

$$A_{[jik]} = \frac{1}{3!}(A_{jik} + A_{kji} + A_{ikj} - A_{jki} - A_{ijk} - A_{kij}) = -A_{[ijk]} \tag{107}$$

and hence it is anti-symmetric in i and j. The exchange of other indices can be done similarly and hence the symmetric nature of $A_{(ijk)}$ and the anti-symmetric nature of $A_{[ijk]}$ with respect to their other indices can also be verified. The symmetric and anti-symmetric parts, $A_{(ijk)}$ and $A_{[jik]}$, are related by the following equations:

$$3\left(A_{(ijk)} + A_{[jik]}\right) = A_{ijk} + A_{kij} + A_{jki} \tag{108}$$
$$3\left(A_{(ijk)} - A_{[jik]}\right) = A_{ikj} + A_{jik} + A_{kji} \tag{109}$$

which can also be verified by substitution, as done for rank-2. The right hand side of the first equation is the sum of the even permutations of the indices ijk while the right hand side of the second equation is the sum of the odd permutations of these indices.

More generally, a rank-n tensor $A_{i_1\ldots i_n}$ can be symmetrized by:

$$A_{(i_1\ldots i_n)} = \frac{1}{n!}\left(\sum \text{even permutations of } i\text{'s} + \sum \text{odd permutations of } i\text{'s}\right) \tag{110}$$

and anti-symmetrized by:

$$A_{[i_1\ldots i_n]} = \frac{1}{n!}\left(\sum \text{even permutations of } i\text{'s} - \sum \text{odd permutations of } i\text{'s}\right) \tag{111}$$

Similar verifications and relations can be established for the rank-n case following the patterns seen in the above rank-2 and rank-3 cases.

3.1.5 Symmetric and Anti-symmetric Tensors

A tensor of high rank (> 2) may be symmetrized or anti-symmetrized with respect to only some of its indices instead of all of its indices. For example, in the following the tensor A_{ijk} is symmetrized and anti-symmetrized respectively with respect to its first two indices only, that is:

$$A_{(ij)k} = \frac{1}{2}(A_{ijk} + A_{jik}) \tag{112}$$

$$A_{[ij]k} = \frac{1}{2}(A_{ijk} - A_{jik}) \tag{113}$$

The symmetry of $A_{(ij)k}$ and the anti-symmetry of $A_{[ij]k}$ with respect to the ij indices can be verified by exchanging these indices in the above relations as done previously for $A_{(ij)}$ and $A_{[ij]}$. Moreover, the tensor A_{ijk} can be expressed as the sum of these symmetric and anti-symmetric tensors, that is:

$$A_{(ij)k} + A_{[ij]k} = \frac{1}{2}(A_{ijk} + A_{jik}) + \frac{1}{2}(A_{ijk} - A_{jik}) = A_{ijk} \tag{114}$$

A tensor is described as totally symmetric *iff* it is symmetric with respect to all of its indices, that is:

$$A_{i_1...i_n} = A_{(i_1...i_n)} \tag{115}$$

and hence its anti-symmetric part is zero, i.e. $A_{[i_1...i_n]} = 0$. Similarly, a tensor is described as totally anti-symmetric *iff* it is anti-symmetric in all of its indices, that is:

$$A_{i_1...i_n} = A_{[i_1...i_n]} \tag{116}$$

and hence its symmetric part is zero, i.e. $A_{(i_1...i_n)} = 0$.

For a totally anti-symmetric tensor, non-zero entries can occur only when all the indices are different. This is because an exchange of two identical indices, which identifies identical entries, should change the sign due to the anti-symmetry and no number can be equal to its negation except zero. Therefore, if the tensor A_{ijk} is totally anti-symmetric then only its entries of the form A_{ijk}, where $i \neq j \neq k$, are not identically zero while all the other entries (i.e. all those of the from A_{iij}, A_{iji}, A_{jii} and A_{iii} where $i \neq j$ with no sum on i) vanish identically. In fact, this also applies to partially anti-symmetric tensors where the entries corresponding to identical anti-symmetric indices should vanish identically. Hence, if A_{ijkl} is anti-symmetric in its first two indices and in its last two indices then only its entries with $i \neq j$ and $k \neq l$ do not vanish identically while all its entries with $i = j$ or $k = l$ (or both) are identically zero.

It should be remarked that the indices whose exchange defines the symmetry and anti-symmetry relations should be of the same variance type, i.e. both upper or both lower. Hence, a tensor A_i^j is not symmetric if the components satisfy the relation $A_i^j = A_j^i$ or anti-symmetric if the components satisfy the relation $A_i^j = -A_j^i$. This should be obvious by considering that the covariant and contravariant indices correspond to different basis sets (i.e. contravariant and covariant), as explained in § 3.1.1.

Another important remark is that the symmetry and anti-symmetry characteristic of a tensor is invariant under coordinate transformations. Hence, a symmetric/anti-symmetric tensor in one coordinate system is symmetric/anti-symmetric in all other coordinate systems. This is because if we label the tensor as **A** and its transformation by symmetry or anti-symmetry as **B**, then an equality of the two tensors based on their symmetric or anti-symmetric nature (i.e. $A_{ij} = B_{ji}$ or $A_{ij} = -B_{ji}$) can be expressed, using an algebraic transformation, as a zero tensor (i.e. $A_{ij} - B_{ji} = 0$ or $A_{ij} + B_{ji} = 0$) which is invariant under coordinate transformations as stated earlier (see § 3.1.4). Similarly, a tensor which is neither symmetric nor anti-symmetric in one coordinate system should remain so in all other coordinate systems obtained by permissible transformations. This is based on the previous statement because if it is symmetric or anti-symmetric in one coordinate system it should be symmetric or anti-symmetric in all other coordinate systems according to the previous statement. In brief, the characteristic of a tensor as being symmetric or anti-symmetric or neither is invariant under permissible coordinate transformations.

Finally, for a symmetric tensor A_{ij} and an anti-symmetric tensor B^{ij} (or the other way around) we have the following useful and widely used identity:

$$A_{ij}B^{ij} = 0 \qquad (117)$$

This is because an exchange of the i and j indices will change the sign of one tensor only, which is the anti-symmetric tensor, and this will change the sign of the term in the summation resulting in having a sum of terms which is identically zero due to the fact that each term in the sum has its own negation. This also includes the zero components of the anti-symmetric tensor where the terms containing these components are zero.

3.1.6 General and Affine Tensors

Affine tensors are tensors that correspond to admissible linear coordinate transformations (see § 2.3.4) from an original rectangular system of coordinates, while general tensors correspond to any type of admissible coordinate transformations. These categories are part of the terminology of tensor calculus which we use in this book and hence they do not have a particular significance.

3.2 Tensor Operations

In this section, we briefly examine the main elementary tensor operations, which are mostly of algebraic nature, that permeate tensor algebra and calculus. There are various operations that can be performed on tensors to produce other tensors in general. Examples of these operations are addition/subtraction, multiplication by a scalar (rank-0 tensor), multiplication of tensors (each of rank > 0), contraction and permutation. Some of these operations, such as addition and multiplication, involve more than one tensor while others, such as permutation, are performed on a single tensor. Contraction can involve one tensor or two tensors.

Before we start our investigation, we should remark that the last subsection (i.e. § 3.2.7), which is about the quotient rule of tensors, is added to the present section of tensor operations because this section is the most appropriate place for it in the present book considering the dependency of the definition of this rule on other tensor operations; otherwise the subsection is not about a tensor operation in the same sense as the operations presented in the other subsections. Another remark is that in tensor algebra division is allowed only for scalars, and hence if the components of an indexed tensor should appear in a denominator, the tensor should be redefined to avoid this. For example, if A_i represents the components of a tensor and we should have $\frac{1}{A_i}$ as the components of another tensor, then we introduce another tensor, say B_i, whose components are the reciprocals of A_i (i.e. $B_i = \frac{1}{A_i}$) and use B_i instead of $\frac{1}{A_i}$. The purpose of this change of notation is to avoid confusion and facilitate the manipulation of tensors according to the familiar rules of indices.

3.2.1 Addition and Subtraction

Tensors of the same rank and type can be added algebraically to produce a tensor of the same rank and type, e.g.

$$a = b + c \qquad\qquad A_i = B_i - C_i \qquad\qquad A^i_j = B^i_j + C^i_j \qquad (118)$$

In this operation the entries of the two tensors are added algebraically componentwise. For example, in the second equality of the last equation the C_1 component of the tensor **C** is subtracted from the B_1 component of the tensor **B** to produce the A_1 component of the tensor **A**, while in the third equality the B^1_2 component of the tensor **B** is added to the C^1_2 component of the tensor **C** to produce the A^1_2 component of the tensor **A**.

We note that "type" in the above statement refers to variance type (covariant, contravariant, mixed) and true/pseudo type as well as other qualifications to which the tensors participating in an addition or subtraction operation should match such as having the same weight if they are relative tensors, as outlined previously (refer for example to § 3.1.3). The added/subtracted terms should also have the same indicial structure with regard to their free indices, as explained in § 1.2 in the context of the rules governing the indices of tensor expressions and equalities. Hence, the tensors A^i_{jk} and B^j_{ik} cannot be added or subtracted although they are of the same rank and type, but the tensors A^i_{jk} and B^i_{jk} can be added and subtracted. In brief, the tensors involved in addition and subtraction operations should satisfy all the rules that have been stated previously on the terms of tensor expressions and equalities.

Addition of tensors is associative and commutative, that is:

$$(\mathbf{A} + \mathbf{B}) + \mathbf{C} = \mathbf{A} + (\mathbf{B} + \mathbf{C}) \qquad (119)$$
$$\mathbf{A} + \mathbf{B} = \mathbf{B} + \mathbf{A} \qquad (120)$$

In fact, associativity and commutativity can include subtraction if the minus sign is absorbed in the subtracted tensor; in which case the operation is converted to addition.

3.2.2 Multiplication of Tensor by Scalar

A tensor can be multiplied by a scalar, which should not be zero in general, to produce a tensor of the same variance type, rank and indicial structure, e.g.

$$A_{ik}^{j} = aB_{ik}^{j} \tag{121}$$

where a is a non-zero scalar. As indicated by this equation, multiplying a tensor by a scalar means multiplying each component of the tensor by that scalar. Multiplication by a scalar is commutative and distributive over algebraic sum of tensors, that is:

$$a\mathbf{A} = \mathbf{A}a \tag{122}$$
$$a(\mathbf{A} \pm \mathbf{B}) = a\mathbf{A} \pm a\mathbf{B} \tag{123}$$

where a is a scalar and $\mathbf{A}$ and $\mathbf{B}$ are non-scalar tensors. It is also associative when more than two factors are involved, that is:

$$a(b\mathbf{A}) = (ab)\mathbf{A} \tag{124}$$
$$a(\mathbf{A} \circ \mathbf{B}) = (a\mathbf{A}) \circ \mathbf{B} = \mathbf{A} \circ (a\mathbf{B}) \tag{125}$$

where b is another scalar and $\circ$ represents an outer or inner product operation (see § 3.2.3 and 3.2.5).

3.2.3 Tensor Multiplication

This operation, which can be defined generically as multiplication between two non-scalar tensors, may also be called outer or exterior or direct or dyadic multiplication, although some of these names may be reserved to operations on vectors. On multiplying each component of a tensor of rank r by each component of a tensor of rank k, both of dimension d, a tensor of rank $(r+k)$ with d^{r+k} components is obtained where the variance type of each index (covariant or contravariant) is preserved. More clearly, by multiplying a tensor of type (m_1, n_1, w_1) by a tensor of type (m_2, n_2, w_2) a tensor of type $(m_1+m_2, n_1+n_2, w_1+w_2)$ is obtained. This means that the tensor rank and weight in the outer product operation are additive and the operation conserves the variance type of each index of the tensors involved. Also, the order of the indices in the product should match the order of the indices in the multiplicands, as will be seen in the following examples, since tensor indices refer to specific tensor bases and the multiplication of tensors is not commutative (refer to § 3.1.1).

For example, if $\mathbf{A}$ and $\mathbf{B}$ are covariant tensors of rank-1, then on multiplying $\mathbf{A}$ by $\mathbf{B}$ we obtain a covariant tensor $\mathbf{C}$ of rank-2 and type $(0, 2)$ where the components of $\mathbf{C}$ are given by:

$$C_{ij} = A_i B_j \tag{126}$$

while on multiplying $\mathbf{B}$ by $\mathbf{A}$ we obtain a covariant tensor $\mathbf{D}$ of rank-2 and type $(0, 2)$ where the components of $\mathbf{D}$ are given by:

$$D_{ij} = B_i A_j \tag{127}$$

3.2.3 Tensor Multiplication

Similarly, if **A** is a contravariant tensor of rank-2 and **B** is a covariant tensor of rank-2, then on multiplying **A** by **B** we obtain a mixed tensor **C** of rank-4 and type $(2,2)$ where the components of **C** are given by:

$$C^{ij}{}_{kl} = A^{ij} B_{kl} \qquad (128)$$

while on multiplying **B** by **A** we obtain a mixed tensor **D** of rank-4 and type $(2,2)$ where the components of **D** are given by:

$$D_{ij}{}^{kl} = B_{ij} A^{kl} \qquad (129)$$

In general, the outer product of tensors yields a tensor. In the outer product operation, it is generally understood that all the indices of the involved tensors have the same range (i.e. all the tensors have the same dimension) although this may not always be the case. As indicated before (see § 1.2), there are cases of tensors which are not uniformly dimensioned, and in some cases these tensors may be regarded as the result of an outer product of lower rank tensors.

The direct multiplication of tensors may be marked by the symbol $\otimes$, mostly when using symbolic notation for tensors, e.g. $\mathbf{A} \otimes \mathbf{B}$. However, in the present book no symbol will be used to represent the operation of direct multiplication[15] and hence the operation is symbolized by putting the symbols of the tensors side by side, e.g. **AB** where **A** and **B** are non-scalar tensors. In this regard, the reader should be vigilant to avoid confusion with the operation of matrix multiplication which, according to the notation of matrix algebra, is also symbolized as **AB** where **A** and **B** are matrices of compatible dimensions, since matrix multiplication from tensor perspective is an inner product, rather than an outer product, operation.

The direct multiplication of tensors is not commutative in general as indicated above; however it is distributive with respect to algebraic sum of tensors, that is:

$$\mathbf{AB} \neq \mathbf{BA} \qquad (130)$$

$$\mathbf{A}(\mathbf{B} \pm \mathbf{C}) = \mathbf{AB} \pm \mathbf{AC} \qquad (\mathbf{B} \pm \mathbf{C})\mathbf{A} = \mathbf{BA} \pm \mathbf{CA} \qquad (131)$$

Regarding the associativity of direct multiplication, there are cases in which this operation is not associative according to the research literature of tensor calculus.

As indicated before, the rank-2 tensor constructed by the direct multiplication of two vectors is commonly called dyad. More generally, tensors may be expressed as an outer product of vectors where the rank of the resultant product is equal to the number of the vectors involved, i.e. 2 for dyads, 3 for triads and so on. However, not every tensor can be synthesized as a product of lower rank tensors. Multiplication of a tensor by a scalar (refer to § 3.2.2) may be regarded as a special case of direct multiplication since it is a tensor multiplication operation with one of the tensors involved being of rank-0 although the definition of direct multiplication seems to restrict this operation to non-scalar tensors, as stated above.

---

[15] We mean "specifically" because we use the symbol ∘ to represent general tensor multiplication which includes inner and outer tensor products.

3.2.4 Contraction

The contraction operation of a tensor of rank > 1 is to make two free indices identical, by unifying their symbols, followed by performing summation over these repeated indices, e.g.

$$A_i^j \xrightarrow{\text{contraction}} A_i^i \tag{132}$$

$$A_{il}^{jk} \xrightarrow{\text{contraction on } jl} A_{im}^{mk} \tag{133}$$

Accordingly, contraction results in a reduction of the rank by 2 since it requires the annihilation of two free indices by the summation operation. Therefore, the contraction of a rank-2 tensor results in a scalar, the contraction of a rank-3 tensor results in a vector, the contraction of a rank-4 tensor results in a rank-2 tensor, and so on. Contraction can also take place between two tensors as part of an inner product operation, as will be explained in § 3.2.5, although the contraction in this case is actually done on the tensor resulting from the outer product operation that underlies the inner product.

For general coordinate systems, the pair of contracted indices should be different in their variance type, i.e. one upper and one lower. Hence, contraction of a mixed tensor of type (m, n) will in general produce a tensor of type $(m - 1, n - 1)$. A tensor of type (p, q) can therefore have $p \times q$ possible contractions, i.e. one contraction for each combination of lower and upper indices. As indicated before, there is no difference between the covariant and contravariant types in orthonormal Cartesian systems and hence contraction can take place between any pair of indices. Accordingly, a rank-n tensor in orthonormal Cartesian systems can have $\frac{n(n-1)}{2}$ possible individual contraction operations where each one of the n indices can be contracted with each one of the remaining $(n-1)$ indices and the factor 2 in the denominator represents the fact that the contraction operation is independent of the order of the two contracted indices since contracting i with j is the same as contracting j with i.[16] We note that conducting a contraction operation on a tensor results into a tensor. Similarly, the application of a contraction operation on a relative tensor (see § 3.1.3) produces a relative tensor of the same weight as the original tensor.

A common example of a contraction operation conducted on a single tensor is the operation of taking the trace of a square matrix, as defined in matrix algebra, by summing its diagonal elements, which can be considered as a contraction operation on the rank-2 tensor represented by the matrix, and hence it yields the trace which is a scalar. Similarly, a well known example of a contraction operation that takes place between two tensors is the dot product operation on vectors which can be considered as a direct multiplication (refer to § 3.2.3) of the two vectors that results in a rank-2 tensor followed by a contraction operation and hence it produces a scalar.

3.2.5 Inner Product

On taking the outer product (refer to § 3.2.3) of two tensors of rank ≥ 1 followed by a contraction (refer to § 3.2.4) on two indices of the product, an inner product of the two

---

[16] This may also be formulated by the rule of combination.

3.2.5 Inner Product

tensors is formed. Hence, if one of the original tensors is of rank-m and the other is of rank-n, the inner product will be of rank-$(m+n-2)$. In the symbolic notation of tensor calculus, the inner product operation is usually symbolized by a single dot between the two tensors, e.g. $\mathbf{A} \cdot \mathbf{B}$, to indicate the contraction operation which follows the outer multiplication.

In general, the inner product is not commutative. When one or both of the tensors involved in the inner product are of rank > 1 then the order of the multiplicands does matter in general, that is:

$$\mathbf{A} \cdot \mathbf{B} \neq \mathbf{B} \cdot \mathbf{A} \tag{134}$$

However, the inner product operation is distributive with respect to the algebraic sum of tensors, that is:

$$\mathbf{A} \cdot (\mathbf{B} \pm \mathbf{C}) = \mathbf{A} \cdot \mathbf{B} \pm \mathbf{A} \cdot \mathbf{C} \qquad (\mathbf{B} \pm \mathbf{C}) \cdot \mathbf{A} = \mathbf{B} \cdot \mathbf{A} \pm \mathbf{C} \cdot \mathbf{A} \tag{135}$$

As indicated before (see § 3.2.4), the dot product of two vectors is an example of the inner product of tensors, i.e. it is an inner product of two rank-1 tensors to produce a rank-0 tensor. For example, if $\mathbf{a}$ is a covariant vector and $\mathbf{b}$ is a contravariant vector, then their dot product can be depicted as follows:

$$[\mathbf{ab}]_i^{\;j} = a_i b^j \quad \xrightarrow{\text{contraction}} \quad \mathbf{a} \cdot \mathbf{b} = a_i b^i \tag{136}$$

Another common example, from linear algebra, of inner product is the multiplication of a matrix representing a rank-2 tensor, by a vector, which is a rank-1 tensor, to produce a vector. For example, if $\mathbf{A}$ is a rank-2 covariant tensor and $\mathbf{b}$ is a contravariant vector, then their inner product can be depicted, according to tensor calculus, as follows:

$$[\mathbf{Ab}]_{ij}^{\;\;k} = A_{ij} b^k \quad \xrightarrow{\text{contraction on } jk} \quad [\mathbf{A} \cdot \mathbf{b}]_i = A_{ij} b^j \tag{137}$$

This operation is equivalent to the aforementioned operation of multiplying a matrix by a vector as defined in linear algebra. We note that we are using here the symbolic notation of tensor calculus, rather than the matrix notation, in writing $\mathbf{Ab}$ and $\mathbf{A} \cdot \mathbf{b}$ to represent, respectively, the outer and inner products. In matrix notation, $\mathbf{Ab}$ is used to represent the product of a matrix by a vector which is an inner product according to the terminology of tensor calculus. The multiplication of two $n \times n$ matrices, as defined in linear algebra, to produce another $n \times n$ matrix is another example of inner product. In this operation, each one of the matrices involved in the multiplication, as well as the product itself, represents a rank-2 tensor.

For tensors whose outer product produces a tensor of rank > 2 and type (m,n) where $m,n > 0$, various contraction operations between different pairs of indices of opposite variance type can occur and hence more than one inner product, which are different in general, can be defined. Moreover, when the outer product produces a tensor of rank > 3 and type (m,n) where $m,n > 1$, more than one contraction operation can take place simultaneously. Based on what we have seen earlier (refer to § 3.2.3 and 3.2.4), the outer product of a tensor of type (p,q) by a tensor of type (s,t) produces a tensor of type

$(p+s, q+t)$ and hence $(p+s) \times (q+t)$ individual inner product operations can take place, i.e. one inner product for each combination of lower and upper indices. We note that for orthonormal Cartesian systems the variance type is irrelevant and hence a rank-n tensor can have $\frac{n(n-1)}{2}$ individual inner product operations which is the number of possible contraction operations, as explained earlier (see § 3.2.4).

There are more specialized types of inner product; some of these may be defined differently by different authors. For example, a double contraction inner product of two rank-2 tensors, **A** and **B**, may be defined and denoted by double vertically- or horizontally-aligned dots (i.e. **A** : **B** or **A** $\cdot\cdot$ **B**) to indicate double contraction taking place between different pairs of indices. An instance of these specialized types is the inner product with double contraction of two dyads which is commonly defined by:

$$\mathbf{ab} : \mathbf{cd} = (\mathbf{a} \cdot \mathbf{c})(\mathbf{b} \cdot \mathbf{d}) \tag{138}$$

where the single dots in the right hand side of this equation symbolize the conventional dot product of two vectors. The result of this operation is obviously a scalar since it is the product of two scalars, as can be seen from the right hand side of the equation.

Some authors may define a different type of double contraction inner product of two dyads, symbolized by two horizontally-aligned dots, which may be called a transposed contraction. This type of inner product is given by:

$$\mathbf{ab} \cdot\cdot \mathbf{cd} = \mathbf{ab} : \mathbf{dc} = (\mathbf{a} \cdot \mathbf{d})(\mathbf{b} \cdot \mathbf{c}) \tag{139}$$

where the result is also a scalar. The second equality of the last equation is based on Eq. 138. We note that the double inner product operators, i.e. : and $\cdot\cdot$, are defined by some authors opposite to the above definitions (i.e. the other way around) and hence one should be on the lookout for such differences in convention.

For two rank-2 tensors, the aforementioned double contraction inner products are similarly defined as in the case of two dyads, that is:

$$\mathbf{A} : \mathbf{B} = A_{ij} B^{ij} \qquad\qquad \mathbf{A} \cdot\cdot \mathbf{B} = A_{ij} B^{ji} \tag{140}$$

Inner products with higher multiplicities of contraction can be defined similarly, and hence they may be regarded as trivial extensions of the inner products with lower contraction multiplicities. Finally, we note that the inner product of tensors produces a tensor because the inner product is an outer product operation followed by a contraction operation and both of these operations on tensors produce tensors, as stated before (see 3.2.3 and 3.2.4).

3.2.6 Permutation

A tensor may be obtained by exchanging the indices of another tensor. For example, A^i_{kj} is a permutation of the tensor A^i_{jk}. A common example of the permutation operation of tensors is the transposition of a matrix representing a rank-2 tensor since the first and second indices, which represent the rows and columns of the matrix, are exchanged in this operation. It is obvious that tensor permutation applies only to tensors of rank > 1 since no exchange of indices can take place in a scalar with no index or in a vector with a single index. Also, permutation may be restricted to indices of the same variance type. The collection of tensors obtained by permuting the indices of a given tensor may be called isomers.

3.2.7 Tensor Test and Quotient Rule

Sometimes a tensor-like object may be suspected for being a tensor; in such cases a test based on what is called the "quotient rule" can be used to clarify the situation. We should remark that the quotient rule of tensors must not be confused with the quotient rule of differentiation. According to the quotient rule of tensors, if the inner product of a suspected tensor by a known tensor is a tensor then the suspect is a tensor. In more formal terms, if it is not known if **A** is a tensor or not but it is known that **B** and **C** are tensors; moreover it is known that the following relation holds true in all rotated (i.e. properly-transformed) coordinate frames:

$$A_{pq...k...m} B_{ij...k...n} = C_{pq...mij...n} \tag{141}$$

then **A** is a tensor. Here, **A**, **B** and **C** are respectively of ranks m, n and $(m+n-2)$, where the rank of **C** is reduced by 2 due to the contraction on k which can be any index of **A** and **B** independently. We assume, of course, that the rules of contraction of indices, such as being of opposite variance type in the case of non-Cartesian coordinates, are satisfied in this operation. The form given by the above equation is based, for simplicity, on assuming a Cartesian system.

Finally, we should remark that testing a suspected tensor for being a tensor can also be done by employing the first principles through direct application of the transformation rules to see if the alleged tensor satisfies the transformation rules of tensors or not. However, using the quotient rule is generally more convenient and requires less work. Another remark is that the quotient rule of tensors may be considered by some authors as a replacement for the division operation which is not defined for tensors.

3.3 Tensor Representations

So far, we are familiar with the covariant and contravariant (including mixed) representations of tensors. There is still another type of representation, that is the physical representation which is the common one in the scientific applications of tensor calculus such as fluid and continuum mechanics. The introduction and employment of the physical representation of tensors are justified by the fact that the covariant and contravariant basis vectors, as well as the covariant and contravariant components of a vector, do not in general have the same physical dimensions as explained earlier (see § 2.2). Moreover, the basis vectors may not have the same magnitude. This motivates the introduction of a more standard form of vectors by using physical components (which have the same dimensions) with normalized basis vectors (which are dimensionless with unit magnitude) where the metric tensor and the scale factors are employed to facilitate this process. The normalization of the basis vectors is done by dividing each vector by its magnitude. For example, the normalized covariant and contravariant basis vectors of a general coordinate system, $\hat{\mathbf{E}}_i$ and $\hat{\mathbf{E}}^i$, are given by:[17]

$$\hat{\mathbf{E}}_i = \frac{\mathbf{E}_i}{|\mathbf{E}_i|} \qquad\qquad \hat{\mathbf{E}}^i = \frac{\mathbf{E}^i}{|\mathbf{E}^i|} \qquad \text{(no sum on } i\text{)} \qquad (142)$$

which for an orthogonal coordinate system becomes (see Eq. 70):

$$\hat{\mathbf{E}}_i = \frac{\mathbf{E}_i}{\sqrt{g_{ii}}} = \frac{\mathbf{E}_i}{h_i} \qquad\qquad \hat{\mathbf{E}}^i = \frac{\mathbf{E}^i}{\sqrt{g^{ii}}} = h_i \mathbf{E}^i \qquad \text{(no sum on } i\text{)} \qquad (143)$$

where g_{ii} and g^{ii} are the ith diagonal elements of the covariant and contravariant metric tensor respectively and h_i is the scale factor of the ith coordinate as described previously (see § 2.5 and 2.6).

Consequently, if the physical components of a contravariant vector are notated with a hat, then for an orthogonal system we have:

$$\mathbf{A} = A^i \mathbf{E}_i = \hat{A}^i \hat{\mathbf{E}}_i = \hat{A}^i \frac{\mathbf{E}_i}{\sqrt{g_{ii}}} \qquad\Longrightarrow\qquad \hat{A}^i = \sqrt{g_{ii}} A^i = h_i A^i \qquad \text{(no sum)} \quad (144)$$

Similarly, for the covariant form of the vector we have:

$$\mathbf{A} = A_i \mathbf{E}^i = \hat{A}_i \hat{\mathbf{E}}^i = \hat{A}_i \frac{\mathbf{E}^i}{\sqrt{g^{ii}}} \qquad\Longrightarrow\qquad \hat{A}_i = \sqrt{g^{ii}} A_i = \frac{A_i}{\sqrt{g_{ii}}} = \frac{A_i}{h_i} \qquad \text{(no sum)} \quad (145)$$

These definitions and processes can be easily extended to tensors of higher ranks as we will see next.

The physical components of higher rank tensors are similarly defined as for rank-1 tensors by considering the basis vectors of the coordinate system of the space where similar simplifications apply to orthogonal coordinate systems. For example, for a rank-2 tensor $\mathbf{A}$ in an orthogonal coordinate system, the physical components can be represented by:

$$\hat{A}_{ij} = \frac{A_{ij}}{h_i h_j} \qquad \text{(no sum on } i \text{ or } j \text{, with basis } \hat{\mathbf{E}}^i \hat{\mathbf{E}}^j\text{)} \qquad (146)$$

$$\hat{A}^{ij} = h_i h_j A^{ij} \qquad \text{(no sum on } i \text{ or } j \text{, with basis } \hat{\mathbf{E}}_i \hat{\mathbf{E}}_j\text{)} \qquad (147)$$

$$\hat{A}^i_{\ j} = \frac{h_i A^i_{\ j}}{h_j} \qquad \text{(no sum on } i \text{ or } j \text{, with basis } \hat{\mathbf{E}}_i \hat{\mathbf{E}}^j\text{)} \qquad (148)$$

On generalizing the above pattern, we conclude that the physical components of a tensor of type (m, n) in an orthogonal coordinate system are given by:

$$\hat{A}^{a_1 \ldots a_m}_{b_1 \ldots b_n} = \frac{h_{a_1} \ldots h_{a_m}}{h_{b_1} \ldots h_{b_n}} A^{a_1 \ldots a_m}_{b_1 \ldots b_n} \qquad \text{(no sum on any index)} \qquad (149)$$

---

[17] We note that the factors in the denominators are scalars and hence the rules of indices are not violated.

where the symbols and basis tensors are defined similarly following the above examples.

As a consequence of the last statements, in a space with a well defined metric any tensor can be expressed in covariant or contravariant (including mixed) or physical forms using different sets of basis tensors. Moreover, these forms can be transformed from each other by using the raising and lowering operators and scale factors. As before, for orthonormal Cartesian systems the covariant, contravariant and physical components are the same where the Kronecker delta is the metric tensor. This is because the covariant, contravariant and physical basis vectors are identical in these systems.

More generally, for orthogonal coordinate systems the two sets of normalized covariant and contravariant basis vectors are identical as established earlier because the corresponding vectors of each basis set are in the same direction (see § 2.4 and 2.6), and hence the physical components corresponding to the covariant and contravariant components are identical as well. Consequently, for orthogonal coordinate systems with orthonormal basis vectors, the covariant, contravariant and physical components are identical because the normalized bases corresponding to these three forms are identical.

The physical components of a tensor may be represented by the symbol of the tensor with subscripts denoting the coordinates of the employed coordinate system. For instance, if **A** is a vector in a 3D space with contravariant components A^i or covariant components A_i, its physical components in Cartesian, cylindrical, spherical and general curvilinear systems may be denoted by (A_x, A_y, A_z), (A_ρ, A_ϕ, A_z), (A_r, A_θ, A_ϕ) and (A_u, A_v, A_w) respectively. For consistency and dimensional homogeneity, the tensors in scientific applications are commonly represented by their physical components with a set of normalized basis vectors. The invariance of the tensor form then guarantees that the same tensor formulation is valid regardless of any particular coordinate system where standard tensor transformations can be used to convert from one form to another without affecting the validity and invariance of the formulation.

3.4 Exercises and Revision

3.1 Define "covariant" and "contravariant" tensors from the perspective of their notation and their transformation rules.

3.2 Write the transformation relations for covariant and contravariant vectors and for covariant, contravariant and mixed rank-2 tensors between different coordinate systems.

3.3 State the practical rules for writing the transformation relations of tensors between different coordinate systems.

3.4 What are the raising and lowering operators and how they provide the link between the covariant and contravariant types?

3.5 **A** is a tensor of type (m, n) and **B** is a tensor of type (p, q, w). What this means? Write these tensors in their indicial form.

3.6 Write the following equations in full tensor notation and explain their significance:

$$\mathbf{E}_i = \frac{\partial \mathbf{r}}{\partial u^i} \qquad \mathbf{E}^i = \nabla u^i$$

3.4 Exercises and Revision

3.7 Write the orthonormalized form of the basis vectors in a 2D general coordinate system. Verify that these vectors are actually orthonormal.

3.8 Why the following relations are labeled as the reciprocity relations?

$$\mathbf{E}_i \cdot \mathbf{E}^j = \delta_i^j \qquad\qquad \mathbf{E}^i \cdot \mathbf{E}_j = \delta^i{}_j$$

3.9 The components of the tensors **A**, **B** and **C** are given by: $A_{ik}^{\cdot\cdot\ j}$, $B^{jn}_{\cdot\cdot\ mq}$ and $C_{k\cdot i}^{\ l}$. Write these tensors in their full notation that includes their basis tensors.

3.10 **A**, **B** and **C** are tensors of rank-2, rank-3 and rank-4 respectively in a given coordinate system. Write the components of these tensors with respect to the following basis tensors: , $\mathbf{E}^i\mathbf{E}^n$, $\mathbf{E}_i\mathbf{E}_k\mathbf{E}_m$ and $\mathbf{E}_j\mathbf{E}^i\mathbf{E}_k\mathbf{E}^n$.

3.11 What "dyad" means? Write all the nine unit dyads associated with the double directions of rank-2 tensors in a 3D space with a rectangular Cartesian coordinate system (i.e. $\mathbf{e}_1\mathbf{e}_1 \cdots \mathbf{e}_3\mathbf{e}_3$).

3.12 Make a simple sketch of the nine dyads of exercise 3.11.

3.13 Compare true and pseudo vectors making a clear distinction between the two with a simple illustrating plot. Generalize this to tensors of any rank.

3.14 Justify the following statement: "The terms of consistent tensor expressions and equations should be uniform in their true and pseudo type".

3.15 What is the curl of a pseudo vector from the perspective of true/pseudo qualification?

3.16 Define absolute and relative tensors stating any necessary mathematical relations.

3.17 What is the weight of the product of **A** and **B** where **A** is a tensor of type $(1, 2, 2)$ and **B** is a tensor of type $(0, 3, -1)$?

3.18 Show that the determinant of a rank-2 absolute tensor **A** is a relative scalar and find the weight in the case of **A** being covariant and in the case of **A** being contravariant.

3.19 Why the tensor terms of tensor expressions and equalities should have the same weight?

3.20 What "isotropic" and "anisotropic" tensor mean?

3.21 Give an example of an isotropic rank-2 tensor and another example of an anisotropic rank-3 tensor.

3.22 What is the significance of the fact that the zero tensor of all ranks and all dimensions is isotropic with regard to the invariance of tensors under coordinate transformations?

3.23 Define "symmetric" and "anti-symmetric" tensor. Why scalars and vectors are not qualified to be symmetric or anti-symmetric?

3.24 Write the symmetric and anti-symmetric parts of the tensor A^{ij}.

3.25 Write the symmetrization and anti-symmetrization formulae for a rank-n tensor $A^{i_1\ldots i_n}$.

3.26 Symmetrize and anti-symmetrize the tensor A^{ijkl} with respect to its second and fourth indices.

3.27 Write the two mathematical conditions for a rank-n tensor $A^{i_1\ldots i_n}$ to be totally symmetric and totally anti-symmetric.

3.28 The tensor A_{ijk} is totally symmetric. How many distinct components it has in a 3D space?

3.29 The tensor B_{ijk} is totally anti-symmetric. How many identically vanishing components it has in a 3D space? How many distinct non-identically vanishing components it has in a 3D space?

3.30 Give numeric or symbolic examples of a rank-2 symmetric tensor and a rank-2 skew-symmetric tensor in a 4D space. Count the number of independent non-zero components in each case.

3.31 Write the formula for the number of independent components of a rank-2 symmetric tensor, and the formula for the number of independent non-zero components of a rank-2 anti-symmetric tensor in nD space.

3.32 Explain why the entries corresponding to identical anti-symmetric indices should vanish identically.

3.33 Why the indices whose exchange defines the symmetry and anti-symmetry relations should be of the same variance type?

3.34 Discuss the significance of the fact that the symmetry and anti-symmetry characteristic of a tensor is invariant under coordinate transformations and link this to the invariance of the zero tensor.

3.35 Verify the relation $A_{ij}B^{ij} = 0$, where A_{ij} is a symmetric tensor and B^{ij} is an anti-symmetric tensor, by writing the sum in full assuming a 3D space.

3.36 Classify the common tensor operations with respect to the number of tensors involved in these operations.

3.37 Which of the following operations are commutative, associative or distributive when these properties apply: algebraic addition, algebraic subtraction, multiplication by a scalar, outer multiplication, and inner multiplication?

3.38 For question 3.37, write all the required mathematical relations that describe those properties.

3.39 The tensors involved in tensor addition, subtraction or equality should be compatible in their types. Give all the details about these "types".

3.40 What is the meaning of multiplying a tensor by a scalar in terms of the components of the tensor?

3.41 A tensor of type (m_1, n_1, w_1) is multiplied by another tensor of type (m_2, n_2, w_2). What is the type, the rank and the weight of the product?

3.42 We have two tensors: $\mathbf{A} = A_{ij}\mathbf{E}^i\mathbf{E}^j$ and $\mathbf{B} = B^{kl}\mathbf{E}_k\mathbf{E}_l$. We also have $\mathbf{C} = \mathbf{AB}$ and $\mathbf{D} = \mathbf{BA}$. Use the properties of tensor operations to obtain the full expression of $\mathbf{C}$ and $\mathbf{D}$ in terms of their components and basis tensors (i.e. $\mathbf{C} = \mathbf{AB} = \cdots$ etc.).

3.43 Explain why tensor multiplication, unlike ordinary multiplication of scalars, is not commutative considering the basis tensors to which the tensors are referred.

3.44 The direct product of vectors $\mathbf{a}$ and $\mathbf{b}$ is $\mathbf{ab}$. Edit the following equation by adding a simple notation to make it correct without changing the order: $\mathbf{ab} = \mathbf{ba}$.

3.45 What is the difference in notation between matrix multiplication and tensor multiplication of two tensors, $\mathbf{A}$ and $\mathbf{B}$, when we write $\mathbf{AB}$?

3.46 Define the contraction operation of tensors. Why this operation cannot be conducted on scalars and vectors?

3.47 In reference to general coordinate systems, a single contraction operation is conducted on a tensor of type (m, n, w) where $m, n > 0$. What is the rank, the type and the weight of the contracted tensor?

3.4 Exercises and Revision

3.48 What is the condition that should be satisfied by the two tensor indices involved in a contraction operation assuming a general coordinate system? What about tensors in orthonormal Cartesian systems?

3.49 How many individual contraction operations can take place in a tensor of type (m, n, w) in a general coordinate system? Explain why.

3.50 How many individual contraction operations can take place in a rank-n tensor in an orthonormal Cartesian coordinate system? Explain why.

3.51 List all the possible single contraction operations that can take place in the tensor A^{ijk}_{lm}.

3.52 List all the possible double contraction operations that can take place in the tensor A^{ij}_{kmn}.

3.53 Give examples of contraction operation from matrix algebra.

3.54 Show that contracting a rank-n tensor results in a rank-$(n-2)$ tensor.

3.55 Discuss inner multiplication of tensors as an operation composed of two more simple operations.

3.56 Give common examples of inner product operation from linear algebra and vector calculus.

3.57 Why inner product operation is not commutative in general?

3.58 Complete the following equations where **A** and **B** are rank-2 tensors of opposite variance type:

$$\mathbf{A} : \mathbf{B} = ? \qquad \mathbf{A} \cdot\cdot \mathbf{B} = ?$$

3.59 Write $\mathbf{ab} : \mathbf{cd}$ in component form assuming a Cartesian system. Repeat this with $\mathbf{ab} \cdot\cdot \mathbf{cd}$.

3.60 Why the operation of inner multiplication of tensors results in a tensor?

3.61 We have: $\mathbf{A} = A^i \mathbf{E}_i$, $\mathbf{B} = B_j \mathbf{E}^j$ and $\mathbf{C} = A^m_{n} \mathbf{E}_m \mathbf{E}^n$. Find the following tensor products: **AB**, **AC** and **BC**.

3.62 Referring to the tensors in question 3.61, find the following dot products: $\mathbf{B} \cdot \mathbf{B}$, $\mathbf{C} \cdot \mathbf{A}$ and $\mathbf{C} \cdot \mathbf{B}$.

3.63 Define permutation of tensors giving an example of this operation from matrix algebra.

3.64 State the quotient rule of tensors in words and in a formal mathematical form.

3.65 Why the quotient rule is usually used in tensor tests instead of applying the transformation rules?

3.66 Outline the similarities and differences between the three main forms of tensor representation, i.e. covariant, contravariant and physical.

3.67 Define, mathematically, the physical basis vectors, $\hat{\mathbf{E}}_i$ and $\hat{\mathbf{E}}^i$, in terms of the covariant and contravariant basis vectors, $\mathbf{E}_i$ and $\mathbf{E}^i$.

3.68 Correct, if necessary, the following relation: $\hat{A}^{ikn}_{jm} = \frac{h_i h_j h_n}{h_k h_m} A^{ikn}_{jm}$ (no sum on any index) where **A** is a tensor in an orthogonal coordinate system.

3.69 Why the normalized covariant, contravariant and physical basis vectors are identical in orthogonal coordinate systems?

3.70 What is the physical significance of being able to transform one type of tensors to other types as well as transforming between different coordinate systems?

3.71 Why the physical representation of tensors is usually preferred in the scientific appli-

3.4 Exercises and Revision

cations of tensor calculus?

3.72 Give a few common examples of physical representation of tensors in mathematical and scientific applications.

3.73 What is the advantage of representing the physical components of a tensor (e.g. **A**) by the symbol of the tensor with subscripts denoting the coordinates of the employed coordinate system, e.g. (A_r, A_θ, A_ϕ) in spherical coordinate systems?

Chapter 4
Special Tensors

The subject of investigation of this chapter is those tensors that form an essential part of the tensor calculus theory itself, namely the Kronecker, the permutation and the metric tensors. The chapter also includes a section devoted to the generalized Kronecker delta tensor which may be regarded as a bridge between the ordinary Kronecker delta tensor and the permutation tensor. There is also another section in which a number of important tensor identities related to the Kronecker or/and permutation tensors are collected and discussed. Finally, there is a section devoted to some important mathematical definitions and applications in which special tensors are employed.

The Kronecker and permutation tensors are of particular importance in tensor calculus due to their distinctive properties and unique transformation attributes. They are numerical tensors with invariant components in all coordinate systems. They enter in the definition of many mathematical objects in tensor calculus and are used to facilitate the formulation of many tensor identities. Similarly, the metric tensor is one of the most important tensors (and may even be the most important) in tensor calculus and its applications, as will be revealed in § 4.5 and other parts of this book. In fact, it permeates the whole subject of tensor calculus due to its role in characterizing the space and its involvement in essential mathematical definitions, operations and transformations. For example, it enters in the definition of many mathematical concepts, such as curve length and surface area, and facilitates the transformation between covariant and contravariant types of tensors and their basis vectors.

4.1 Kronecker delta Tensor

As indicated above, there are two types of Kronecker delta tensor: ordinary and generalized. In the present section we discuss the ordinary Kronecker delta tensor and in § 4.4 we investigate the generalized Kronecker delta tensor.

The ordinary Kronecker delta tensor, which may also be called the unit tensor, is a rank-2 numeric, absolute, symmetric, constant, isotropic tensor in all dimensions. It is defined in its covariant form as:

$$\delta_{ij} = \begin{cases} 1 & (i = j) \\ 0 & (i \neq j) \end{cases} \qquad (i,j = 1, 2, \ldots n) \qquad (150)$$

where n is the space dimension, and hence it can be considered as the identity tensor or matrix. For example, in a 3D space the Kronecker delta tensor is given by:

$$[\delta_{ij}] = \begin{bmatrix} \delta_{11} & \delta_{12} & \delta_{13} \\ \delta_{21} & \delta_{22} & \delta_{23} \\ \delta_{31} & \delta_{32} & \delta_{33} \end{bmatrix} = \begin{bmatrix} 1 & 0 & 0 \\ 0 & 1 & 0 \\ 0 & 0 & 1 \end{bmatrix} \qquad (151)$$

4.2 Permutation Tensor

The contravariant and mixed forms of the ordinary Kronecker delta tensor, i.e. δ^{ij} and δ^j_i, are similarly defined. Consequently, the numerical values of the components of the covariant, contravariant and mixed types of the Kronecker delta tensor are the same, that is:

$$\delta_{ij} = \delta^{ij} = \delta^i_j = \delta^j_i \qquad (152)$$

and hence they are all defined by Eq. 150. We note that in the last equation (and in any similar equation) the numerical values of the components (not the tensors) are equated and hence this is not a violation of the rules of tensor indices as stated in § 1.2. The tensor is made of components with reference to a set of basis vectors and hence the equality of the components of two tensors of different variance type does not imply the equality of the two tensors since the two basis vector sets to which the two tensors are referred can be different.

From the above definitions, it can be seen that the Kronecker delta tensor is symmetric, that is:

$$\delta_{ij} = \delta_{ji} \qquad\qquad \delta^{ij} = \delta^{ji} \qquad (153)$$

where $i, j = 1, 2, \ldots, n$. Moreover, the tensor is conserved under all proper and improper coordinate transformations where "conserved" means that the tensor keeps the numerical values of its components following a coordinate transformation. Since it is conserved under proper transformations, it is an isotropic tensor. We note that being conserved under all transformations is stronger than being isotropic as the former applies even under improper coordinate transformations while isotropy is restricted, by definition, to proper transformations (see § 3.1.4). We also used "conserved" rather than "invariant" to indicate the preservation of the components and to avoid confusion with form-invariance which is a property that characterizes all tensors.

4.2 Permutation Tensor

The permutation tensor, which is a numeric tensor with constant components, is also known as the Levi-Civita, anti-symmetric and alternating tensor. We note that the "Levi-Civita" label is usually used for the rank-3 permutation tensor. Also some authors distinguish between the permutation tensor and the Levi-Civita tensor even for rank-3. Moreover, some of the common labels and descriptions of the permutation tensor are more specific to rank-3. Hence, differences in conventions, definitions and labels should be considered when reading the literature of tensor calculus related to the permutation tensor.

The permutation tensor has a rank equal to the number of dimensions of the space. Hence, a rank-n permutation tensor has n^n components. This tensor is characterized by the following properties:

1. It is numeric tensor and hence the value of its components are: -1, 1 and 0 in all coordinate systems.
2. The value of any particular component (e.g. ϵ_{312}) of this tensor is the same in any coordinate system and hence it is constant tensor in this sense.

4.2 Permutation Tensor

3. It is relative tensor of weight -1 for its covariant form and $+1$ for its contravariant form.
4. It is isotropic tensor (see § 3.1.4) since its components are conserved under proper transformations.
5. It is totally anti-symmetric in each pair of its indices, i.e. it changes sign on swapping any two of its indices.
6. It is pseudo tensor (see § 3.1.2) since it acquires a minus sign under improper orthogonal transformation of coordinates.
7. The permutation tensor of any rank has only one independent non-vanishing component because all the non-zero components of this tensor are of unity magnitude.
8. The rank-n permutation tensor possesses $n!$ non-zero components which is the number of the non-repetitive permutations of its indices.

The rank-2 permutation tensor ϵ_{ij} in its covariant form is defined by:[18]

$$\epsilon_{12} = 1 \qquad \epsilon_{21} = -1 \qquad \epsilon_{11} = \epsilon_{22} = 0 \qquad (154)$$

Similarly, the rank-3 permutation tensor ϵ_{ijk} in its covariant form is defined by:

$$\epsilon_{ijk} = \begin{cases} 1 & (i,j,k \text{ is even permutation of 1,2,3}) \\ -1 & (i,j,k \text{ is odd permutation of 1,2,3}) \\ 0 & (\text{repeated index}) \end{cases} \qquad (155)$$

The contravariant form of the rank-2 and rank-3 permutation tensors is similarly defined.

Figure 15 is a graphical illustration of the rank-3 permutation tensor ϵ_{ijk} while Figure 16, which may be used as a mnemonic device, demonstrates the cyclic nature of the three even permutations of the indices of the rank-3 permutation tensor and the three odd permutations of these indices assuming no repetition of indices. The three permutations in each case are obtained by starting from a given number in the cycle and rotating in the given sense to obtain the other two numbers in the permutation.

The definition of the rank-n permutation tensor (i.e. $\epsilon_{i_1 i_2 \ldots i_n}$ and $\epsilon^{i_1 i_2 \ldots i_n}$) is similar to the definition of the rank-3 permutation tensor with regard to the repetition in its indices $(i_1, i_2, \cdots, i_n)$ and being even or odd permutations in their correspondence to $(1, 2, \cdots, n)$, that is:

$$\epsilon_{i_1 i_2 \ldots i_n} = \epsilon^{i_1 i_2 \ldots i_n} = \begin{cases} 1 & (i_1, i_2, \ldots, i_n \text{ is even permutation of } 1, 2, \ldots, n) \\ -1 & (i_1, i_2, \ldots, i_n \text{ is odd permutation of } 1, 2, \ldots, n) \\ 0 & (\text{repeated index}) \end{cases} \qquad (156)$$

As stated before, equations like this defines the numeric values of the tensor components and hence they do not violate the rules of indices with regard to their variance type (see § 1.2) or the rules of relative tensors with regard to their weight (see § 3.1.3).

As well as the inductive definition of the permutation tensor (as given by Eqs. 154, 155 and 156), the permutation tensor of any rank can also be defined analytically where the

---

[18] There is no rank-1 permutation tensor as there is no possibility of permutation in a 1D space.

4.2 Permutation Tensor

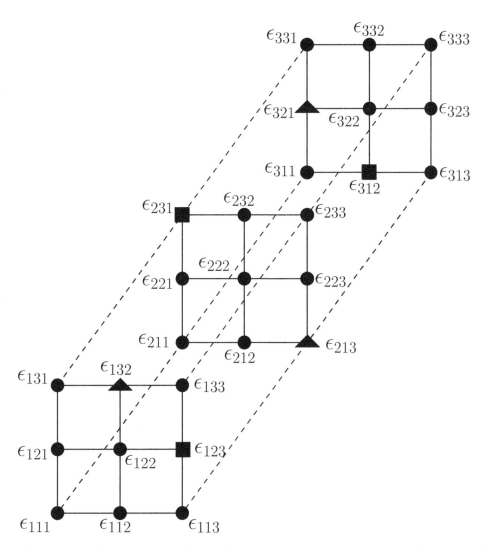

Figure 15: Graphical illustration of the rank-3 permutation tensor ϵ_{ijk} where circular nodes represent 0, square nodes represent 1 and triangular nodes represent -1.

entries of the tensor are calculated from closed form formulae. Accordingly, the values of the entries of the rank-2 permutation tensor can be calculated from the following closed form equation:

$$\epsilon_{ij} = \epsilon^{ij} = (j - i) \tag{157}$$

where each one of the indices i, j ranges over $1, 2$. Similarly, the numerical values of the entries of the rank-3 permutation tensor are given by:

$$\epsilon_{ijk} = \epsilon^{ijk} = \frac{1}{2}(j-i)(k-i)(k-j) \tag{158}$$

where each one of the indices i, j, k ranges over $1, 2, 3$. As for the rank-4 permutation tensor we have:

$$\epsilon_{ijkl} = \epsilon^{ijkl} = \frac{1}{12}(j-i)(k-i)(l-i)(k-j)(l-j)(l-k) \tag{159}$$

4.2 Permutation Tensor

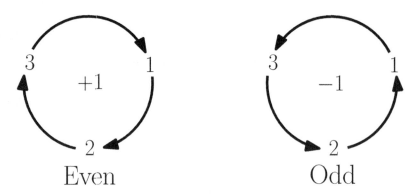

Figure 16: Graphical demonstration of the cyclic nature of the even and odd permutations of the indices of the rank-3 permutation tensor assuming no repetition in indices.

where each one of the indices i, j, k, l ranges over $1, 2, 3, 4$.

More generally, the numerical values of the entries of the rank-n permutation tensor can be obtained from the following formula:

$$\epsilon_{a_1 a_2 \cdots a_n} = \epsilon^{a_1 a_2 \cdots a_n} = \prod_{i=1}^{n-1} \left[\frac{1}{i!} \prod_{j=i+1}^{n} (a_j - a_i) \right] = \frac{1}{S(n-1)} \prod_{1 \leq i < j \leq n} (a_j - a_i) \qquad (160)$$

where each one of the indices $a_1, \cdots, a_n$ ranges over $1, \cdots, n$ and $S(n-1)$ is the super factorial function of the argument $(n-1)$ which is defined by:

$$S(k) = \prod_{i=1}^{k} i! = 1! \cdot 2! \cdot \ldots \cdot k! \qquad (161)$$

A simpler formula for calculating the numerical values of the entries of the rank-n permutation tensor can be obtained from the previous one (Eq. 160) by dropping the magnitude of the multiplication factors and taking their signs only, that is:

$$\epsilon_{a_1 a_2 \cdots a_n} = \epsilon^{a_1 a_2 \cdots a_n} = \prod_{1 \leq i < j \leq n} \mathrm{sgn}\,(a_j - a_i) = \mathrm{sgn}\left(\prod_{1 \leq i < j \leq n} (a_j - a_i)\right) \qquad (162)$$

where $\mathrm{sgn}(k)$ is the sign function of the argument k which is defined by:

$$\mathrm{sgn}(k) = \begin{cases} +1 & (k > 0) \\ -1 & (k < 0) \\ 0 & (k = 0) \end{cases} \qquad (163)$$

The sign function in Eq. 162 can be expressed in a more direct form for the non-vanishing entries of the permutation tensor (which correspond to the non-repetitive permutations) by dividing each argument of the multiplicative factors in Eq. 162 by its absolute value, noting that none of these factors is zero, and hence Eq. 162 becomes:

$$\epsilon_{a_1 a_2 \cdots a_n} = \epsilon^{a_1 a_2 \cdots a_n} = \prod_{1 \leq i < j \leq n} \frac{(a_j - a_i)}{|a_j - a_i|} \qquad (a_j \neq a_i) \qquad (164)$$

4.2 Permutation Tensor

Regarding the vanishing entries, they are easily identified by having repeated indices and hence all the entries of the tensor are easily evaluated.

As stated above, the permutation tensor is totally anti-symmetric (see § 3.1.5) in each pair of its indices, i.e. it changes sign on swapping any two of its indices, that is:

$$\epsilon_{i_1...i_k...i_l...i_n} = -\epsilon_{i_1...i_l...i_k...i_n} \qquad \epsilon^{i_1...i_k...i_l...i_n} = -\epsilon^{i_1...i_l...i_k...i_n} \qquad (165)$$

The reason is that any exchange of two indices requires an even/odd number of single-step shifts to the right of the first index plus an odd/even number of single-step shifts to the left of the second index, so the total number of shifts is odd and hence it is an odd permutation of the original arrangement. This may also be concluded from the closed form formulae (e.g. Eq. 162) where any exchange will lead to an exchange of the indices of a single multiplicative factor which leads to sign change.[19] Also, the permutation tensor is a pseudo tensor since it acquires a minus sign under an improper orthogonal transformation of coordinates, i.e. inversion of axes that changes the system handedness (see § 2.3.1). However, it is an isotropic tensor since it is conserved under proper coordinate transformations.

The permutation tensor may be considered as a contravariant relative tensor of weight +1 or as a covariant relative tensor of weight −1. Since the contravariant and covariant types of the permutation tensor are relative tensors, it is desirable to define absolute covariant and contravariant forms of the permutation tensor. This is done by using the metric tensor, that is:

$$\underline{\epsilon}_{i_1...i_n} = \sqrt{g}\,\epsilon_{i_1...i_n} \qquad \underline{\epsilon}^{i_1...i_n} = \frac{1}{\sqrt{g}}\epsilon^{i_1...i_n} \qquad (166)$$

where the indexed ϵ and $\underline{\epsilon}$ are respectively the relative and absolute permutation tensors of the given type, and g is the determinant of the covariant metric tensor g_{ij}. Referring to Eqs. 63 and 93, we see that the $\underline{\epsilon}$ tensors are absolute with $w = 0$. We note that the contravariant form of the absolute permutation tensor requires a sign function but the details are out of the scope of the present text (see Zwillinger in the References). However, for the rank-3 permutation tensor, which is the one used mostly in the present book, the above expression stands as it is.

The contravariant and covariant types of the rank-3 permutation tensor are linked, through the Kronecker delta tensor, by the following relations:

$$\epsilon^{ijk}\epsilon_{lmk} = \delta^i_l \delta^j_m - \delta^i_m \delta^j_l \qquad (167)$$

$$\epsilon^{ijk}\epsilon_{lmn} = \begin{vmatrix} \delta^i_l & \delta^i_m & \delta^i_n \\ \delta^j_l & \delta^j_m & \delta^j_n \\ \delta^k_l & \delta^k_m & \delta^k_n \end{vmatrix} \qquad (168)$$

---

[19] The non-existence of a mixed type permutation tensor is also related to its totally anti-symmetric nature.

where the indexed δ represent the mixed form of the ordinary Kronecker delta. The last equation can be generalized to nD spaces as follows:

$$\epsilon^{i_1 i_2 \cdots i_n} \epsilon_{j_1 j_2 \cdots j_n} = \begin{vmatrix} \delta^{i_1}_{j_1} & \delta^{i_1}_{j_2} & \cdots & \delta^{i_1}_{j_n} \\ \delta^{i_2}_{j_1} & \delta^{i_2}_{j_2} & \cdots & \delta^{i_2}_{j_n} \\ \vdots & \vdots & \ddots & \vdots \\ \delta^{i_n}_{j_1} & \delta^{i_n}_{j_2} & \cdots & \delta^{i_n}_{j_n} \end{vmatrix} \tag{169}$$

The contravariant and covariant types of the rank-n permutation tensor are also linked by the following relation:

$$\epsilon^{i_1 i_2 \cdots i_n} \epsilon_{i_1 i_2 \cdots i_n} = n! \tag{170}$$

where n is the space dimension. The left hand side of this equation represents the sum of the products of the corresponding entries of the contravariant and covariant types of the permutation tensor.

On comparing Eq. 169 and the upcoming Eq. 199 we obtain the following identity which links the contravariant and covariant types of the permutation tensor to each other and to the generalized Kronecker delta:

$$\epsilon^{i_1 \cdots i_n} \epsilon_{j_1 \cdots j_n} = \delta^{i_1 \cdots i_n}_{j_1 \cdots j_n} \tag{171}$$

More details about these identities will be given in the subsequent sections of this chapter.

4.3 Identities Involving Kronecker or/and Permutation Tensors

4.3.1 Identities Involving Kronecker delta Tensor

When an index of the Kronecker delta tensor is involved in a contraction operation by repeating an index in another tensor in its own term, the effect of this is to replace the shared index in the other tensor by the other index of the Kronecker delta, that is:

$$\delta^j_i A_j = A_i \qquad \delta^i_j A^j = A^i \tag{172}$$

In such cases the Kronecker delta is described as an index replacement or substitution operator. Hence, we have:

$$\delta^j_i \delta^k_j = \delta^k_i \tag{173}$$

Similarly, we have:

$$\delta^j_i \delta^k_j \delta^i_k = \delta^k_i \delta^i_k = \delta^i_i = n \tag{174}$$

where n is the space dimension. The last part of this equation (i.e. $\delta^i_i = n$) can be easily justified by the fact that δ^i_i is the trace of the identity tensor considering the summation convention.

Due to the fact that the coordinates are independent of each other (see § 2.2), we also have the following identity which is based on the well known rules of partial differentiation:

$$\frac{\partial u^i}{\partial u^j} = \partial_j u^i = u^i{}_{,j} = \delta^i_j \tag{175}$$

Hence, in an nD space we obtain the following identity from the identities of Eqs. 175 and 174:
$$\partial_i u^i = \delta_i^i = n \tag{176}$$

Based on the above identities and facts, the following identity can be shown to apply in orthonormal Cartesian coordinate systems:
$$\frac{\partial x_i}{\partial x_j} = \delta_{ij} = \delta_{ji} = \frac{\partial x_j}{\partial x_i} \tag{177}$$

This identity is based on the two facts that the coordinates are independent, and the covariant and contravariant types are the same in orthonormal Cartesian coordinate systems.

Similarly, for a set of orthonormal vectors, such as the basis set of orthonormal Cartesian system, the following identity can be easily proved:
$$\mathbf{e}_i \cdot \mathbf{e}_j = \delta_{ij} \tag{178}$$

where the indexed $\mathbf{e}$ represents the orthonormal vectors. This identity is no more than a symbolic statement of the fact that the vectors in such a set are mutually orthogonal and each one is of unit length. We note that for orthonormal basis sets the covariant and contravariant forms are identical as explained earlier.

Finally, the double inner product of two dyads formed by an orthonormal set of vectors satisfies the following identity (see § 3.2.5):
$$\mathbf{e}_i \mathbf{e}_j : \mathbf{e}_k \mathbf{e}_l = (\mathbf{e}_i \cdot \mathbf{e}_k)(\mathbf{e}_j \cdot \mathbf{e}_l) = \delta_{ik} \delta_{jl} \tag{179}$$

This identity is a combination of Eq. 138 and Eq. 178.

4.3.2 Identities Involving Permutation Tensor

From the definition of the rank-3 permutation tensor, we obtain the following identity which demonstrates the sense of cyclic order of the non-repetitive permutations of this tensor:
$$\epsilon_{ijk} = \epsilon_{kij} = \epsilon_{jki} = -\epsilon_{ikj} = -\epsilon_{jik} = -\epsilon_{kji} \tag{180}$$

This identity is also a demonstration of the fact that the rank-3 permutation tensor is totally anti-symmetric in all of its indices since a shift of any two indices reverses its sign. Moreover, it reflects the fact that this tensor has only one independent non-zero component since any one of the non-zero entries, all of which are given by Eq. 180, can be obtained from any other one of these entries by at most a reversal of sign. This identity also applies to the contravariant form of the permutation tensor.

We also have the following identity for the rank-n permutation tensor:
$$\epsilon^{i_1 i_2 \cdots i_n} \epsilon_{i_1 i_2 \cdots i_n} = n! \tag{181}$$

This identity is based on the fact that the left hand side is the sum of the squares of the epsilon symbol over all the $n!$ non-repetitive permutations of n different indices where the value of epsilon of each one of these permutations is either $+1$ or -1 and hence in both cases their square is $+1$. The repetitive permutations are zero and hence they do not contribute to the sum.

The double inner product of the rank-3 permutation tensor and a symmetric tensor **A** is given by the following identity:

$$\epsilon_{ijk} A^{ij} = \epsilon_{ijk} A^{ik} = \epsilon_{ijk} A^{jk} = 0 \tag{182}$$

This is because an exchange of the two indices of **A** does not affect its value due to the symmetry of **A** whereas a similar exchange of these indices in ϵ_{ijk} results in a sign change; hence each term in the sum has its own negative and therefore the total sum is identically zero. This identity also applies to the contravariant permutation tensor with a covariant symmetric tensor.

Another identity with a trivial outcome that involves the rank-3 permutation tensor and a vector **A** is the following:

$$\epsilon_{ijk} A^i A^j = \epsilon_{ijk} A^i A^k = \epsilon_{ijk} A^j A^k = 0 \tag{183}$$

This can be explained similarly by the fact that, due to the commutativity of ordinary multiplication, an exchange of the indices in A's will not affect the value but a similar exchange in the corresponding indices of ϵ_{ijk} will cause a change in sign; hence each term in the sum has its own negative and therefore the total sum will be zero. This identity also applies to the contravariant permutation tensor with a covariant vector. We also note that the identities of Eqs. 182 and 183 similarly apply to other ranks of the permutation tensor as they are based on the totally anti-symmetric property of this tensor.

Finally, for a set of three orthonormal vectors forming a right handed system, the following identities are satisfied:

$$\mathbf{e}_i \times \mathbf{e}_j = \epsilon_{ijk} \mathbf{e}_k \tag{184}$$
$$\mathbf{e}_i \cdot (\mathbf{e}_j \times \mathbf{e}_k) = \epsilon_{ijk} \tag{185}$$

These identities are based, respectively, on the forthcoming definitions of the cross product (see Eq. 494) and the scalar triple product (see Eq. 495) in tensor notation plus the fact that these vectors are unit vectors.

4.3.3 Identities Involving Kronecker and Permutation Tensors

For the rank-2 permutation tensor, we have the following identity which involves the ordinary Kronecker delta tensor in 2D:

$$\epsilon^{ij} \epsilon_{kl} = \begin{vmatrix} \delta^i_k & \delta^i_l \\ \delta^j_k & \delta^j_l \end{vmatrix} = \delta^i_k \delta^j_l - \delta^i_l \delta^j_k \tag{186}$$

4.3.3 Identities Involving Kronecker and Permutation Tensors

Table 2: Truth table for the identity of Eq. 186.

i	j	k	l	ϵ^{ij}	ϵ_{kl}	$\epsilon^{ij}\epsilon_{kl}$	δ^i_k	δ^j_l	δ^i_l	δ^j_k	$\delta^i_k\delta^j_l - \delta^i_l\delta^j_k$
1	1	1	1	0	0	**0**	1	1	1	1	**0**
1	1	1	2	0	1	**0**	1	0	0	1	**0**
1	1	2	1	0	-1	**0**	0	1	1	0	**0**
1	1	2	2	0	0	**0**	0	0	0	0	**0**
1	2	1	1	1	0	**0**	1	0	1	0	**0**
1	2	1	2	1	1	**1**	1	1	0	0	**1**
1	2	2	1	1	-1	**-1**	0	0	1	1	**-1**
1	2	2	2	1	0	**0**	0	1	0	1	**0**
2	1	1	1	-1	0	**0**	0	1	0	1	**0**
2	1	1	2	-1	1	**-1**	0	0	1	1	**-1**
2	1	2	1	-1	-1	**1**	1	1	0	0	**1**
2	1	2	2	-1	0	**0**	1	0	1	0	**0**
2	2	1	1	0	0	**0**	0	0	0	0	**0**
2	2	1	2	0	1	**0**	0	1	1	0	**0**
2	2	2	1	0	-1	**0**	1	0	0	1	**0**
2	2	2	2	0	0	**0**	1	1	1	1	**0**

This identity can be proved inductively by building a table for the values on the left and right hand sides as the indices are varied, as seen in Table 2. The pattern of the indices in the determinant array of this identity is simple, that is the indices of the first ϵ provide the indices for the rows as the upper indices of the deltas while the indices of the second ϵ provide the indices for the columns as the lower indices of the deltas. In fact, the role of these indices in indexing the rows and columns can be shifted. This can be explained by the fact that the positions of the two epsilons can be exchanged, since ordinary multiplication is commutative, and hence the role of the epsilons in providing the indices for the rows and columns will be shifted. This can also be done by taking the transposition of the array of the determinant, which does not change the value of the determinant since $\det(\mathbf{A}) = \det(\mathbf{A}^T)$.

We note that a table like Table 2 is similar to the truth tables used in verifying Boolean and logic identities, and for this reason we label it as a "truth table". We also note that the method of building a "truth table" like Table 2 can also be used for proving other similar identities. The main advantage of this method is that it is easy and straightforward while its main disadvantage is that it is lengthy and hence it may not be feasible for some messy identities. Moreover, it can provide proofs for special cases but it is not general with respect to proving similar identities in a general nD space where n is variable although it may be used as a part of an inductive proof. Another advantage of this method is that it lends itself to programming and hence it is ideal for use in computing. We note that the subject matter in these tables is the tensor components.

Another useful identity involving the rank-2 permutation tensor with the Kronecker delta

4.3.3 Identities Involving Kronecker and Permutation Tensors

tensor in 2D is the following:
$$\epsilon^{il}\epsilon_{kl} = \delta^i_k \tag{187}$$

This can be obtained from the identity of Eq. 186 by replacing j with l followed by minimal algebraic manipulations using tensor calculus rules, that is:

$$\begin{aligned}
\epsilon^{il}\epsilon_{kl} &= \delta^i_k\delta^l_l - \delta^i_l\delta^l_k & \text{(Eq. 186 with } j \to l\text{)} \\
&= 2\delta^i_k - \delta^i_l\delta^l_k & \text{(Eq. 174)} \\
&= 2\delta^i_k - \delta^i_k & \text{(Eq. 173)} \\
&= \delta^i_k
\end{aligned} \tag{188}$$

Similarly, we have the following identity which correlates the rank-3 permutation tensor to the Kronecker delta tensor in 3D:

$$\epsilon^{ijk}\epsilon_{lmn} = \begin{vmatrix} \delta^i_l & \delta^i_m & \delta^i_n \\ \delta^j_l & \delta^j_m & \delta^j_n \\ \delta^k_l & \delta^k_m & \delta^k_n \end{vmatrix} = \left(\delta^i_l\delta^j_m\delta^k_n + \delta^i_m\delta^j_n\delta^k_l + \delta^i_n\delta^j_l\delta^k_m\right) - \left(\delta^i_l\delta^j_n\delta^k_m + \delta^i_m\delta^j_l\delta^k_n + \delta^i_n\delta^j_m\delta^k_l\right) \tag{189}$$

Again, the indices in the determinant of this identity follow the same pattern as that of Eq. 186. Another pattern can also be seen in the six terms on the right where the three upper indices of all terms are ijk while the three lower indices in the positive terms are the even permutations of lmn and the three lower indices in the negative terms are the odd permutations of lmn. This identity can also be established by a truth table similar to Table 2.

More generally, the determinantal form of Eqs. 186 and 189, which link the rank-2 and rank-3 permutation tensors to the ordinary Kronecker delta tensors in 2D and 3D spaces, can be extended to link the rank-n permutation tensor to the ordinary Kronecker delta tensor in an nD space, that is:

$$\epsilon^{i_1 i_2 \cdots i_n}\epsilon_{j_1 j_2 \cdots j_n} = \begin{vmatrix} \delta^{i_1}_{j_1} & \delta^{i_1}_{j_2} & \cdots & \delta^{i_1}_{j_n} \\ \delta^{i_2}_{j_1} & \delta^{i_2}_{j_2} & \cdots & \delta^{i_2}_{j_n} \\ \vdots & \vdots & \ddots & \vdots \\ \delta^{i_n}_{j_1} & \delta^{i_n}_{j_2} & \cdots & \delta^{i_n}_{j_n} \end{vmatrix} \tag{190}$$

Again, the pattern of the indices in the determinant of this identity in their relation to the indices of the two epsilons follow the same rules as those of Eqs. 186 and 189. Moreover, the proofs of Eqs. 186 and 189 may be extended to Eq. 190 by induction.

Another useful identity in this category is the following:

$$\epsilon^{ijk}\epsilon_{lmk} = \begin{vmatrix} \delta^i_l & \delta^i_m \\ \delta^j_l & \delta^j_m \end{vmatrix} = \delta^i_l\delta^j_m - \delta^i_m\delta^j_l \tag{191}$$

This identity can be obtained from the identity of Eq. 189 by replacing n with k, that is:

$$\epsilon^{ijk}\epsilon_{lmk} = \delta^i_l\delta^j_m\delta^k_k + \delta^i_m\delta^j_k\delta^k_l + \delta^i_k\delta^j_l\delta^k_m - \delta^i_l\delta^j_k\delta^k_m - \delta^i_m\delta^j_l\delta^k_k - \delta^i_k\delta^j_m\delta^k_l \tag{192}$$

4.3.3 Identities Involving Kronecker and Permutation Tensors

$$
\begin{aligned}
&= 3\delta_l^i \delta_m^j + \delta_m^i \delta_l^j + \delta_m^i \delta_l^j - \delta_l^i \delta_m^j - 3\delta_m^i \delta_l^j - \delta_l^j \delta_m^j \\
&= \delta_l^i \delta_m^j - \delta_m^i \delta_l^j \\
&= \begin{vmatrix} \delta_l^i & \delta_m^i \\ \delta_l^j & \delta_m^j \end{vmatrix}
\end{aligned}
$$

The pattern of the indices in this identity (Eq. 191) is as before if we exclude the repetitive index k.

The identity of Eq. 191, which may be called the epsilon-delta identity or the contracted epsilon identity or the Levi-Civita identity, is very useful in manipulating and simplifying tensor expressions and proving vector and tensor identities; examples of which will be seen in § 7.1.5. We note that the determinantal form, seen in the middle equality of Eq. 191, can be considered as a mnemonic device for this identity where the first and second indices of the first ϵ index the rows while the first and second indices of the second ϵ index the columns, as given above. In fact, the determinantal form in all the above equations which are given in this form is a mnemonic device for all these equations, and not only Eq. 191, where the expanded form, if needed, can be easily obtained from the determinant which can be easily built following the simple pattern of indices, as explained above.

Other common identities in this category are:

$$\epsilon^{ijk}\epsilon_{ljk} = 2\delta_l^i \tag{193}$$

$$\epsilon^{ijk}\epsilon_{ijk} = 2\delta_i^i = 2 \times 3 = 3! = 6 \tag{194}$$

The first of these identities can be obtained from Eq. 191 with the replacement of m with j followed by some basic tensor manipulation, that is:

$$
\begin{aligned}
\epsilon^{ijk}\epsilon_{ljk} &= \delta_l^i \delta_j^j - \delta_j^i \delta_l^j \quad &&\text{(Eq. 191 with } m \to j\text{)} \tag{195}\\
&= 3\delta_l^i - \delta_l^i \quad &&\text{(Eqs. 174 and 173)}\\
&= 2\delta_l^i
\end{aligned}
$$

while the second of these identities can be obtained from the first by replacing l with i and applying the summation convention in 3D on the right hand side, i.e. using Eq. 174. The second identity is, in fact, an instance of Eq. 181 for a 3D space.

Another common identity that involves the rank-3 permutation tensor with the ordinary Kronecker delta in a 3D space is the following:

$$\epsilon_{ijk}\delta_1^i \delta_2^j \delta_3^k = \epsilon_{123} = 1 \tag{196}$$

This identity is based in its first part on the use of the ordinary Kronecker delta as an index replacement operator (Eq. 172), where each one of the deltas replaces an index in the permutation tensor, and is based in its second part on the definition of the permutation tensor (Eq. 155).

Finally, the following identity can also be obtained from the definition of the rank-3 permutation tensor (Eq. 155) and the use of the ordinary Kronecker delta as an index replacement operator (Eq. 172):

$$\epsilon_{ijk}\delta_j^i = \epsilon_{ijk}\delta_k^i = \epsilon_{ijk}\delta_i^j = \epsilon_{ijk}\delta_k^j = \epsilon_{ijk}\delta_i^k = \epsilon_{ijk}\delta_j^k = 0 \tag{197}$$

This is because on replacing one of the indices of the permutation tensor it will have two identical indices and hence it is zero, e.g. $\epsilon_{ijk}\delta^i_j = \epsilon_{jjk} = 0$. We note that identities like Eqs. 196 and 197 also apply to the opposite variance type. Also, Eqs. 196 and 197 apply to the permutation tensor of other ranks with some simple modifications.

4.4 Generalized Kronecker delta Tensor

The generalized Kronecker delta tensor in an nD space is an absolute rank-$2n$ tensor of type (n, n) which is normally defined inductively by:

$$\delta^{i_1 \ldots i_n}_{j_1 \ldots j_n} = \begin{cases} 1 & (j_1 \ldots j_n \text{ is even permutation of } i_1 \ldots i_n) \\ -1 & (j_1 \ldots j_n \text{ is odd permutation of } i_1 \ldots i_n) \\ 0 & (\text{repeated } j\text{'s}) \end{cases} \quad (198)$$

It can also be defined analytically by the following $n \times n$ determinant:

$$\delta^{i_1 \ldots i_n}_{j_1 \ldots j_n} = \begin{vmatrix} \delta^{i_1}_{j_1} & \delta^{i_1}_{j_2} & \cdots & \delta^{i_1}_{j_n} \\ \delta^{i_2}_{j_1} & \delta^{i_2}_{j_2} & \cdots & \delta^{i_2}_{j_n} \\ \vdots & \vdots & \ddots & \vdots \\ \delta^{i_n}_{j_1} & \delta^{i_n}_{j_2} & \cdots & \delta^{i_n}_{j_n} \end{vmatrix} \quad (199)$$

where the δ^i_j entries in the determinant are the ordinary Kronecker deltas as defined previously (see § 4.1). In this equation, the pattern of the indices in the generalized Kronecker delta symbol $\delta^{i_1 \ldots i_n}_{j_1 \ldots j_n}$ in connection to the indices in the determinant is similar to the previous patterns seen in § 4.3.3, that is the upper indices in $\delta^{i_1 \ldots i_n}_{j_1 \ldots j_n}$ provide the upper indices in the ordinary deltas by indexing the rows of the determinant, while the lower indices in $\delta^{i_1 \ldots i_n}_{j_1 \ldots j_n}$ provide the lower indices in the ordinary deltas by indexing the columns of the determinant.

From the previous identities, it can be shown that:

$$\epsilon^{i_1 \ldots i_n} \epsilon_{j_1 \ldots j_n} = \begin{vmatrix} \delta^{i_1}_{j_1} & \delta^{i_1}_{j_2} & \cdots & \delta^{i_1}_{j_n} \\ \delta^{i_2}_{j_1} & \delta^{i_2}_{j_2} & \cdots & \delta^{i_2}_{j_n} \\ \vdots & \vdots & \ddots & \vdots \\ \delta^{i_n}_{j_1} & \delta^{i_n}_{j_2} & \cdots & \delta^{i_n}_{j_n} \end{vmatrix} \quad (200)$$

Now, on comparing the last equation with the definition of the generalized Kronecker delta, i.e. Eq. 199, we conclude that:

$$\epsilon^{i_1 \ldots i_n} \epsilon_{j_1 \ldots j_n} = \delta^{i_1 \ldots i_n}_{j_1 \ldots j_n} \quad (201)$$

Based on Eq. 201, the generalized Kronecker delta is the result of multiplying two relative tensors one of weight $w = +1$ and the other of weight $w = -1$ and hence the generalized Kronecker delta has a weight of $w = 0$. This shows that the generalized Kronecker delta is an absolute tensor, as stated above. We remark that the multiplication of relative tensors

4.4 Generalized Kronecker delta Tensor

produces a tensor whose weight is the sum of the weights of the original tensors (see § 3.1.3).

From Eq. 201, we can see that the relation between the rank-n permutation tensor in its covariant and contravariant forms and the generalized Kronecker delta in an nD space is given by:

$$\epsilon_{i_1\ldots i_n} = \epsilon^{1\ldots n}\epsilon_{i_1\ldots i_n} = \delta^{1\ldots n}_{i_1\ldots i_n} \qquad (202)$$

$$\epsilon^{i_1\ldots i_n} = \epsilon^{i_1\ldots i_n}\epsilon_{1\ldots n} = \delta^{i_1\ldots i_n}_{1\ldots n} \qquad (203)$$

where the first of these equations is obtained from Eq. 201 by substituting $(1\ldots n)$ for $(i_1\ldots i_n)$ in the two sides with relabeling j as i and noting that $\epsilon^{1\ldots n} = 1$, while the second equation is obtained from Eq. 201 by substituting $(1\ldots n)$ for $(j_1\ldots j_n)$ in the two sides and noting that $\epsilon_{1\ldots n} = 1$.

Based on Eqs. 202 and 203, the permutation tensor can be considered as an instance of the generalized Kronecker delta. Consequently, the rank-n permutation tensor can be written as an $n \times n$ determinant consisting of the ordinary Kronecker deltas (Eq. 199). Moreover, Eqs. 202 and 203 can provide another definition for the permutation tensor in its covariant and contravariant forms, in addition to the previous inductive and analytic definitions of this tensor which are given by Eqs. 156 and 160.

The 3D generalized Kronecker delta may be symbolized by δ^{ijk}_{lmn}. If we replace n with k and use the determinantal definition of the generalized Kronecker delta of Eq. 199 followed by conducting a few basic algebraic manipulations using some of the above tensor identities, we obtain:

$$\delta^{ijk}_{lmk} = \begin{vmatrix} \delta^i_l & \delta^i_m & \delta^i_k \\ \delta^j_l & \delta^j_m & \delta^j_k \\ \delta^k_l & \delta^k_m & \delta^k_k \end{vmatrix} \qquad \text{(Eq. 199)} \qquad (204)$$

$$= \delta^i_l\left(\delta^j_m\delta^k_k - \delta^j_k\delta^k_m\right) + \delta^i_m\left(\delta^j_k\delta^k_l - \delta^j_l\delta^k_k\right) + \delta^i_k\left(\delta^j_l\delta^k_m - \delta^j_m\delta^k_l\right)$$

$$= \delta^i_l\delta^j_m\delta^k_k - \delta^i_l\delta^j_k\delta^k_m + \delta^i_m\delta^j_k\delta^k_l - \delta^i_m\delta^j_l\delta^k_k + \delta^i_k\delta^j_l\delta^k_m - \delta^i_k\delta^j_m\delta^k_l$$

$$= \delta^i_l\delta^j_m\delta^k_k - \delta^i_l\delta^j_m + \delta^i_m\delta^j_l - \delta^i_m\delta^j_l\delta^k_k + \delta^i_m\delta^j_l - \delta^i_l\delta^j_m \qquad \text{(Eq. 173)}$$

$$= 3\delta^i_l\delta^j_m - \delta^i_l\delta^j_m + \delta^i_m\delta^j_l - 3\delta^i_m\delta^j_l + \delta^i_m\delta^j_l - \delta^i_l\delta^j_m \qquad \text{(Eq. 174)}$$

$$= \delta^i_l\delta^j_m - \delta^i_m\delta^j_l$$

$$= \begin{vmatrix} \delta^i_l & \delta^i_m \\ \delta^j_l & \delta^j_m \end{vmatrix}$$

$$= \delta^{ij}_{lm} \qquad \text{(Eq. 199)}$$

that is:

$$\delta^{ijk}_{lmk} = \delta^{ij}_{lm} \qquad (205)$$

Similar identities can be obtained from contracting two corresponding indices of the nD generalized Kronecker delta to obtain $(n-1)$D generalized Kronecker delta, e.g.

$$\delta^{ij}_{lj} = \delta^i_l \qquad (206)$$

which can be simply verified following the former example, that is:

$$\delta^{ij}_{lj} = \begin{vmatrix} \delta^i_l & \delta^i_j \\ \delta^j_l & \delta^j_j \end{vmatrix} = \delta^i_l \delta^j_j - \delta^i_j \delta^j_l = 2\delta^i_l - \delta^i_l = \delta^i_l \qquad (207)$$

Returning to the widely used epsilon-delta identity of Eq. 191, if we consider Eq. 205 plus the above identities which correlate the permutation tensor, the generalized Kronecker delta tensor and the ordinary Kronecker delta tensor, then an identity equivalent to Eq. 191 that involves only the generalized and ordinary Kronecker deltas can be obtained, that is:[20]

$$\epsilon^{ijk} \epsilon_{lmk} = \delta^{ijk}_{lmk} \qquad \text{(Eq. 201)} \qquad (208)$$
$$= \delta^{ij}_{lm} \qquad \text{(Eq. 205)}$$
$$= \begin{vmatrix} \delta^i_l & \delta^i_m \\ \delta^j_l & \delta^j_m \end{vmatrix} \qquad \text{(Eq. 199)}$$
$$= \delta^i_l \delta^j_m - \delta^i_m \delta^j_l$$

This means that the following relation:

$$\delta^{ijk}_{lmk} = \delta^{ij}_{lm} = \delta^i_l \delta^j_m - \delta^i_m \delta^j_l \qquad (209)$$

which is no more than the definition of the generalized Kronecker delta of Eq. 199 is another form of the epsilon-delta identity. The pattern of the indices on the right hand side in relation to the indices of the generalized Kronecker delta is very simple, that is we take first the corresponding upper and lower indices followed by the diagonally crossed indices (i.e. || − ×). It is worth noting that the epsilon-delta identity (Eqs. 191 and 209) can also be expressed in a more general form by employing the metric tensor with the absolute permutation tensor, that is:

$$g^{ij} \underline{\epsilon}_{ikl} \underline{\epsilon}_{jmn} = g_{km} g_{ln} - g_{kn} g_{lm} \qquad (210)$$

Other identities involving the permutation tensor and the ordinary Kronecker delta tensor can also be formulated in terms of the generalized Kronecker delta tensor.

4.5 Metric Tensor

The metric tensor, which may also be called the fundamental tensor, is a rank-2 symmetric absolute non-singular tensor, where "non-singular" means invertible and hence its determinant does not vanish at any point in the space. The metric tensor is one of the most important special tensors in tensor calculus, if not the most important of all. Its versatile usage and functionalities permeate the whole discipline of tensor calculus and its applications. One of the main objectives of the metric is to generalize the concept of distance to general coordinate systems and hence maintain the invariance of distance

---

[20] In fact, this is a derivation of Eq. 191.

4.5 Metric Tensor

in different coordinate systems, as will be explained next. This tensor is also used to raise and lower indices and thus facilitate the transformation between the covariant and contravariant types. As a tensor, the metric has significance regardless of any coordinate system although it requires a coordinate system to be represented in a specific form (see § 2.7). So, in principle the coordinate system and the space metric are independent entities.

In an orthonormal Cartesian coordinate system of an nD space the length of infinitesimal element of arc, ds, connecting two neighboring points in space, one with coordinates x_i and the other with coordinates $x_i + dx_i$ $(i = 1, \cdots, n)$, is given by:

$$(ds)^2 = dx_i dx_i = \delta_{ij} dx_i dx_j \qquad (211)$$

This definition of distance is the key to introducing a rank-2 tensor, g_{ij}, called the metric tensor which, for a general coordinate system, is defined by:

$$(ds)^2 = g_{ij} du^i du^j \qquad (212)$$

where the indexed u represents general coordinates. The metric tensor in the last equation is of covariant form. The metric tensor has also a contravariant form which is notated with g^{ij}. It is common to reserve the term "metric tensor" to the covariant form and call the contravariant form, which is its inverse, the associate or conjugate or reciprocal metric tensor.

The components of the metric tensor in its covariant and contravariant forms are closely related to the basis vectors, that is:

$$g_{ij} = \mathbf{E}_i \cdot \mathbf{E}_j \qquad (213)$$
$$g^{ij} = \mathbf{E}^i \cdot \mathbf{E}^j \qquad (214)$$

where the indexed $\mathbf{E}$ are the covariant and contravariant basis vectors as defined previously (see § 2.6). Because of these relations, the vectors $\mathbf{E}_i$ and $\mathbf{E}^i$ may be denoted by $\mathbf{g}_i$ and $\mathbf{g}^i$ respectively which is more suggestive of their relation to the metric tensor. Similarly, the mixed type metric tensor is given by:

$$g^i{}_j = \mathbf{E}^i \cdot \mathbf{E}_j = \delta^i{}_j \qquad\qquad g_i{}^j = \mathbf{E}_i \cdot \mathbf{E}^j = \delta_i{}^j \qquad (215)$$

and hence it is the identity tensor. These equalities, which may be described as the reciprocity relations, represent the fact that the covariant and contravariant basis vectors are reciprocal sets.

As a consequence, the covariant metric tensor is given, in full tensor notation, by:

$$g_{ij} = \frac{\partial x^k}{\partial u^i} \frac{\partial x^k}{\partial u^j} \qquad (216)$$

where

$$x^k = x^k \left(u^1, \ldots, u^n \right) \qquad (k = 1, \ldots, n) \qquad (217)$$

4.5 Metric Tensor

are independent coordinates in an nD space with a rectangular Cartesian system, and u^i and u^j ($i, j = 1, \ldots, n$) are independent general coordinates. Similarly, for the contravariant metric tensor we have:

$$g^{ij} = \frac{\partial u^i}{\partial x^k} \frac{\partial u^j}{\partial x^k} \tag{218}$$

As stated already, the basis vectors, whether covariant or contravariant, in general coordinate systems are not necessarily mutually orthogonal and hence the metric tensor is not diagonal in general since the dot products given by Eqs. 213 and 214 (or Eqs. 216 and 218) are not necessarily zero when $i \ne j$. Moreover, since those basis vectors vary in general in their magnitude and relative orientations and they are not necessarily of unit length, the entries of the metric tensor, including the diagonal elements, are not necessarily of unit magnitude. Also, the entries of the metric tensor, including the diagonal elements, can be positive or negative.[21] However, since the dot product of vectors is a commutative operation, the metric tensor is necessarily symmetric. We note that the mixed type metric tensor is diagonal (or in fact the unity tensor) because the covariant and contravariant basis vector sets are reciprocal systems (see Eq. 215). This applies in all coordinate systems.

As indicated above, the covariant and contravariant forms of the metric tensor are inverses of each other and hence we have the following relations:

$$[g_{ij}] = [g^{ij}]^{-1} \qquad\qquad [g^{ij}] = [g_{ij}]^{-1} \tag{219}$$

Hence:

$$g^{ik} g_{kj} = \delta^i{}_j \qquad\qquad g_{ik} g^{kj} = \delta_i{}^j \tag{220}$$

where these equations can be seen to represent matrix multiplication (row×column). A result that can be obtained from the previous statements plus Eqs. 138, 213 and 214 is that:

$$\mathbf{E}^i \mathbf{E}_j : \mathbf{E}^j \mathbf{E}_k = \left(\mathbf{E}^i \cdot \mathbf{E}^j\right)\left(\mathbf{E}_j \cdot \mathbf{E}_k\right) = g^{ij} g_{jk} = \delta^i{}_k \tag{221}$$

$$\mathbf{E}_i \mathbf{E}^j : \mathbf{E}_j \mathbf{E}^k = \left(\mathbf{E}_i \cdot \mathbf{E}_j\right)\left(\mathbf{E}^j \cdot \mathbf{E}^k\right) = g_{ij} g^{jk} = \delta_i{}^k \tag{222}$$

Since the metric tensor has an inverse, it should be non-singular and hence its determinant, which in general is a function of coordinates like the metric tensor itself, should not vanish at any point in the space, that is:

$$g(u^1, \ldots, u^n) = \det(g_{ij}) \ne 0 \tag{223}$$

From the previous statements, it may be concluded that the metric tensor is in fact a transformation of the Kronecker delta tensor in its different variance types from an orthonormal Cartesian coordinate system to a general coordinate system, that is (see Eqs. 73-75):

$$g_{ij} = \frac{\partial x^k}{\partial u^i} \frac{\partial x^l}{\partial u^j} \delta_{kl} = \frac{\partial x^k}{\partial u^i} \frac{\partial x^k}{\partial u^j} = \mathbf{E}_i \cdot \mathbf{E}_j \qquad \text{(covariant)} \tag{224}$$

---

[21] The diagonal entries can be negative when the coordinates are imaginary (see § 2.2.3 and Eq. 241).

4.5 Metric Tensor

$$g^{ij} = \frac{\partial u^i}{\partial x^k}\frac{\partial u^j}{\partial x^l}\delta^{kl} = \frac{\partial u^i}{\partial x^k}\frac{\partial u^j}{\partial x^k} = \mathbf{E}^i \cdot \mathbf{E}^j \qquad \text{(contravariant)} \qquad (225)$$

$$g^i_j = \frac{\partial u^i}{\partial x^k}\frac{\partial x^l}{\partial u^j}\delta^k_l = \frac{\partial u^i}{\partial x^k}\frac{\partial x^k}{\partial u^j} = \mathbf{E}^i \cdot \mathbf{E}_j \qquad \text{(mixed)} \qquad (226)$$

As stated above, the metric tensor is symmetric in its two indices, that is:

$$g_{ij} = g_{ji} \qquad\qquad g^{ij} = g^{ji} \qquad (227)$$

This can be easily explained by the commutativity of the dot product of vectors in reference to the above equations involving the dot product of the basis vectors (Eqs. 213 and 214).

Because of the relations:

$$A^i = \mathbf{A} \cdot \mathbf{E}^i = A_j \mathbf{E}^j \cdot \mathbf{E}^i = A_j g^{ji} \qquad (228)$$

$$A_i = \mathbf{A} \cdot \mathbf{E}_i = A^j \mathbf{E}_j \cdot \mathbf{E}_i = A^j g_{ji} \qquad (229)$$

the metric tensor is used as an operator for raising and lowering indices and hence facilitating the transformation between the covariant and contravariant types of vectors. By a similar argument, the above can be generalized where the contravariant metric tensor is used for raising covariant indices of covariant and mixed tensors and the covariant metric tensor is used for lowering contravariant indices of contravariant and mixed tensors of any rank, e.g.

$$A^i_{\ k} = g^{ij} A_{jk} \qquad\qquad A_i^{\ kl} = g_{ij} A^{jkl} \qquad (230)$$

Consequently, any tensor in a Riemannian space with well-defined metric can be cast into covariant or contravariant or mixed forms where for the mixed form the rank should be > 1. We note that in the operations of raising and lowering of indices the metric tensor acts, like the Kronecker delta tensor, as an index replacement operator beside its action in shifting the index position.

In this context, it should be emphasized that the order of the raised and lowered indices is important and hence:

$$g^{ik} A_{jk} = A_j^{\ i} \qquad\text{and}\qquad g^{ik} A_{kj} = A^i_{\ j} \qquad (231)$$

are different in general. A dot may be used to indicate the original position of the shifted index and hence the order of the indices is recorded, e.g. $A_j^{\ i}$ and $A^i_{\ j}$ for the above examples respectively. Because raising and lowering of indices is a reversible process, keeping a record of the original position of the shifted indices will facilitate the reversal if needed. We note that dots may also be inserted in the symbols of mixed tensors to remove any ambiguity about the order of the indices even without the action of the raising and lowering operators (refer to § 1.2).

For a space with a coordinate system in which the metric tensor can be cast into a diagonal form with all the diagonal entries being of unity magnitude (i.e. ± 1) the space and the metric are called flat. For example, in 2D manifolds a plane surface is a flat space since it can be coordinated by an orthonormal 2D Cartesian system which results

4.5 Metric Tensor

into a diagonal unity metric tensor since the basis vectors of this system are mutually perpendicular constant vectors and each is of unit length. On the other hand, an ellipsoid is not a flat space (i.e. it is curved) because due to its intrinsic curvature it is not possible to find a coordinate system for this type of surface whose basis vectors produce a diagonal metric tensor with constant diagonal elements of unity magnitude. In this context we note that a Riemannian metric, g_{ij}, in a particular coordinate system is a Euclidean metric if it can be transformed to the identity tensor, δ_{ij}, by a permissible coordinate transformation.

If g and $\bar{g}$ are the determinants of the covariant metric tensor in unbarred and barred coordinate systems respectively, i.e. $g = \det(g_{ij})$ and $\bar{g} = \det(\bar{g}_{ij})$, then we have:

$$\bar{g} = J^2 g \qquad\qquad \sqrt{\bar{g}} = J\sqrt{g} \tag{232}$$

where $J \left(= \left|\frac{\partial u}{\partial \bar{u}}\right|\right)$ is the Jacobian of the transformation between the unbarred and barred systems. Consequently, the determinant of the covariant metric tensor and its square root are relative scalars of weight $+2$ and $+1$ respectively (see § 3.1.3).

A "conjugate" or "associate" tensor of a tensor in a metric space is a tensor obtained by inner product multiplication, once or more, of the original tensor by the covariant or contravariant forms of the metric tensor. All tensors associated with a particular tensor through the metric tensor represent the same tensor but in different base systems since the association is no more than raising or lowering indices by the metric tensor which is equivalent to a representation of the components of the tensor relative to different basis sets.

A sufficient and necessary condition for the components of the metric tensor to be constants in a given coordinate system is that the Christoffel symbols of the first or second kind vanish identically (refer to 5.1). This may be concluded from Eqs. 307 and 308. The metric tensor behaves as a constant with respect to covariant and absolute differentiation (see § 5.2 and § 5.3). Hence, in all coordinate systems the covariant and absolute derivatives of the metric tensor are zero. Accordingly, the covariant and absolute derivative operators bypass the metric tensor in differentiating inner and outer products of tensors involving the metric tensor.

In orthogonal coordinate systems in an nD space the metric tensor in its covariant and contravariant forms is diagonal with non-vanishing diagonal elements, that is:

$$g_{ij} = 0 \qquad\qquad g^{ij} = 0 \qquad\qquad (i \neq j) \tag{233}$$

$$g_{ii} \neq 0 \qquad\qquad g^{ii} \neq 0 \qquad\qquad (\text{no sum on } i) \tag{234}$$

Moreover, we have:

$$g_{ii} = (h_i)^2 = \frac{1}{g^{ii}} \qquad\qquad (\text{no sum on } i) \tag{235}$$

$$\det(g_{ij}) = g = g_{11} g_{22} \ldots g_{nn} = \prod_i (h_i)^2 \tag{236}$$

$$\det(g^{ij}) = \frac{1}{g} = \frac{1}{g_{11} g_{22} \ldots g_{nn}} = \left[\prod_i (h_i)^2\right]^{-1} \tag{237}$$

4.5 Metric Tensor

where h_i ($= |\mathbf{E}_i|$) are the scale factors, as described previously in § 2.5 and 2.6.

As indicated before, for orthonormal Cartesian coordinate systems in a 3D space, the metric tensor is given in its covariant and contravariant forms by the 3×3 unit matrix, that is:

$$[g_{ij}] = [\delta_{ij}] = \begin{bmatrix} 1 & 0 & 0 \\ 0 & 1 & 0 \\ 0 & 0 & 1 \end{bmatrix} = [\delta^{ij}] = [g^{ij}] \tag{238}$$

For cylindrical coordinate systems of 3D spaces identified by the coordinates (ρ, ϕ, z), the metric tensor is given in its covariant and contravariant forms by:

$$[g_{ij}] = \begin{bmatrix} 1 & 0 & 0 \\ 0 & \rho^2 & 0 \\ 0 & 0 & 1 \end{bmatrix} \qquad [g^{ij}] = \begin{bmatrix} 1 & 0 & 0 \\ 0 & \frac{1}{\rho^2} & 0 \\ 0 & 0 & 1 \end{bmatrix} \tag{239}$$

while for spherical coordinate systems of 3D spaces identified by the coordinates (r, θ, ϕ), the metric tensor is given in its covariant and contravariant forms by:

$$[g_{ij}] = \begin{bmatrix} 1 & 0 & 0 \\ 0 & r^2 & 0 \\ 0 & 0 & r^2 \sin^2 \theta \end{bmatrix} \qquad [g^{ij}] = \begin{bmatrix} 1 & 0 & 0 \\ 0 & \frac{1}{r^2} & 0 \\ 0 & 0 & \frac{1}{r^2 \sin^2 \theta} \end{bmatrix} \tag{240}$$

As seen in Eqs. 238-240, all these metric tensors are diagonal since all these coordinate systems are orthogonal. We also notice that all the corresponding diagonal elements of the covariant and contravariant types are reciprocals of each other. This can be easily explained by the fact that these two types are inverses of each other, moreover the inverse of an invertible diagonal matrix is a diagonal matrix obtained by taking the reciprocal of the corresponding diagonal elements of the original matrix, as it is known from linear algebra. We also see that the diagonal elements are the squares of the scale factors of these systems or their reciprocals (refer to Table 1).

The Minkowski metric, which is the metric tensor of the 4D space-time of the mechanics of Lorentz transformations, is given by one of the following two forms:

$$[g_{ij}] = [g^{ij}] = \begin{bmatrix} 1 & 0 & 0 & 0 \\ 0 & -1 & 0 & 0 \\ 0 & 0 & -1 & 0 \\ 0 & 0 & 0 & -1 \end{bmatrix} \qquad [g_{ij}] = [g^{ij}] = \begin{bmatrix} -1 & 0 & 0 & 0 \\ 0 & 1 & 0 & 0 \\ 0 & 0 & 1 & 0 \\ 0 & 0 & 0 & 1 \end{bmatrix} \tag{241}$$

Consequently, the length of line element ds can be imaginary (see Eqs. 36 and 212). As seen, this metric belongs to a flat space since it is diagonal with all the diagonal entries being ± 1.

The partial derivatives of the components of the covariant and contravariant metric tensor satisfy the following identities:

$$\partial_k g_{ij} = -g_{mj} g_{ni} \partial_k g^{nm} \tag{242}$$

$$\partial_k g^{ij} = -g^{mj} g^{in} \partial_k g_{nm} \tag{243}$$

4.6 Definitions Involving Special Tensors

A related formula for the partial derivatives of the components of the covariant and contravariant metric tensor is given by:

$$g_{im}\partial_k g^{mj} = -g^{mj}\partial_k g_{im} \tag{244}$$

This relation can be obtained by partial differentiation of the relation $g_{im}g^{mj} = \delta_i^j$ (Eq. 220) with respect to the kth coordinate using the product rule, that is:

$$\partial_k\left(g_{im}g^{mj}\right) = g_{im}\partial_k g^{mj} + g^{mj}\partial_k g_{im} = \partial_k \delta_i^j = 0 \quad \Longrightarrow \quad g_{im}\partial_k g^{mj} = -g^{mj}\partial_k g_{im} \tag{245}$$

The last step in the differentiation (i.e. $\partial_k \delta_i^j = 0$) is justified by the fact that the components of the Kronecker delta tensor are constants.

In fact, Eq. 242 can be obtained form Eq. 244 by relabeling m as n and multiplying both sides of Eq. 244 with g_{mj} followed by contraction and exchanging the labels of j and m, that is:

$$g^{mj}\partial_k g_{im} = -g_{im}\partial_k g^{mj} \qquad \text{(Eq. 244)} \tag{246}$$

$$g^{nj}\partial_k g_{in} = -g_{in}\partial_k g^{nj} \qquad (m \to n) \tag{247}$$

$$g_{mj}g^{nj}\partial_k g_{in} = -g_{mj}g_{in}\partial_k g^{nj} \qquad (\times g_{mj}) \tag{248}$$

$$\delta_m^n \partial_k g_{in} = -g_{mj}g_{in}\partial_k g^{nj} \qquad \text{(Eq. 220)} \tag{249}$$

$$\partial_k g_{im} = -g_{mj}g_{in}\partial_k g^{nj} \qquad \text{(Eq. 172)} \tag{250}$$

$$\partial_k g_{ij} = -g_{mj}g_{ni}\partial_k g^{nm} \qquad (m \leftrightarrow j) \tag{251}$$

Similarly, Eq. 243 can be obtained form Eq. 244 by multiplying both sides of Eq. 244 with g^{in} followed by contraction and exchanging the labels of i and n, that is:

$$g_{im}\partial_k g^{mj} = -g^{mj}\partial_k g_{im} \qquad \text{(Eq. 244)} \tag{252}$$

$$g^{in}g_{im}\partial_k g^{mj} = -g^{in}g^{mj}\partial_k g_{im} \qquad (\times g^{in}) \tag{253}$$

$$\delta_m^n \partial_k g^{mj} = -g^{in}g^{mj}\partial_k g_{im} \qquad \text{(Eq. 220)} \tag{254}$$

$$\partial_k g^{nj} = -g^{in}g^{mj}\partial_k g_{im} \qquad \text{(Eq. 172)} \tag{255}$$

$$\partial_k g^{ij} = -g^{mj}g^{in}\partial_k g_{nm} \qquad (n \leftrightarrow i) \tag{256}$$

4.6 Definitions Involving Special Tensors

In the following subsections, we investigate a number of mathematical objects and operations whose definitions and applications are dependent on the above described special tensors, particularly the permutation and metric tensors. These are just a few examples of tensor definitions involving the special tensors and hence they are not comprehensive in any way.

4.6.1 Dot Product

The dot product of two basis vectors in general coordinate systems was given earlier in § 4.5 (see Eqs. 213-215). This will be used in the present subsection to develop expressions for the dot product of vectors and tensors in general.

The dot product of two vectors, **A** and **B**, in general coordinate systems using their covariant and contravariant forms, as well as opposite forms, is given by:

$$\mathbf{A} \cdot \mathbf{B} = A_i \mathbf{E}^i \cdot B_j \mathbf{E}^j = A_i B_j \mathbf{E}^i \cdot \mathbf{E}^j = g^{ij} A_i B_j = A^j B_j = A_i B^i \tag{257}$$

$$\mathbf{A} \cdot \mathbf{B} = A^i \mathbf{E}_i \cdot B^j \mathbf{E}_j = A^i B^j \mathbf{E}_i \cdot \mathbf{E}_j = g_{ij} A^i B^j = A_j B^j = A^i B_i \tag{258}$$

$$\mathbf{A} \cdot \mathbf{B} = A_i \mathbf{E}^i \cdot B^j \mathbf{E}_j = A_i B^j \mathbf{E}^i \cdot \mathbf{E}_j = \delta^i_j A_i B^j = A_j B^j \tag{259}$$

$$\mathbf{A} \cdot \mathbf{B} = A^i \mathbf{E}_i \cdot B_j \mathbf{E}^j = A^i B_j \mathbf{E}_i \cdot \mathbf{E}^j = \delta^j_i A^i B_j = A^i B_i \tag{260}$$

In brief, the dot product of two vectors is the dot product of their two basis vectors multiplied algebraically by the algebraic product of their components. Because the dot product of the basis vectors is a metric tensor, the metric tensor will act on the components by raising or lowering the index of one component or by replacing the index of a component, as seen in the above equations.

The dot product operations outlined in the previous paragraph can be easily extended to tensors of higher ranks where the covariant and contravariant forms of the components and basis vectors are treated in a similar manner to the above examples to obtain the dot product. For instance, the dot product of a rank-2 tensor of contravariant components A^{ij} and a vector of covariant components B_k is given by:

$$\mathbf{A} \cdot \mathbf{B} = \left(A^{ij} \mathbf{E}_i \mathbf{E}_j \right) \cdot \left(B_k \mathbf{E}^k \right) = A^{ij} B_k \left(\mathbf{E}_i \mathbf{E}_j \cdot \mathbf{E}^k \right) = A^{ij} B_k \mathbf{E}_i \delta^k_j = A^{ij} B_j \mathbf{E}_i \tag{261}$$

that is, the ith component of this product, which is a contravariant vector, is:

$$[\mathbf{A} \cdot \mathbf{B}]^i = A^{ij} B_j \tag{262}$$

From the previous statements, we conclude that the dot product in general coordinate systems occurs between two vectors of opposite variance type. Therefore, to obtain the dot product of two vectors of the same variance type, one of the vectors should be converted to the opposite type by the raising or lowering operator, followed by the inner product operation. This can be generalized to the dot product of higher-rank tensors where the two contracted indices of the dot product should be of opposite variance type and hence the index shifting operator in the form of the metric tensor should be used, if necessary, to achieve this. We note that the generalized dot product of two tensors is invariant under permissible coordinate transformations. We also note that the variance type of the tensors involved in an inner product operation is irrelevant for orthonormal Cartesian systems, as explained before.

4.6.2 Magnitude of Vector

The magnitude of a contravariant vector **A** is given by:

$$|\mathbf{A}| = \sqrt{\mathbf{A} \cdot \mathbf{A}} = \sqrt{(A^i \mathbf{E}_i) \cdot (A^j \mathbf{E}_j)} = \sqrt{(\mathbf{E}_i \cdot \mathbf{E}_j) A^i A^j} = \sqrt{g_{ij} A^i A^j} = \sqrt{A_j A^j} = \sqrt{A^i A_i} \tag{263}$$

where Eqs. 213 and 229 are used. A similar expression can be obtained for the covariant form of the vector, that is:

$$|\mathbf{A}| = \sqrt{\mathbf{A} \cdot \mathbf{A}} = \sqrt{(A_i \mathbf{E}^i) \cdot (A_j \mathbf{E}^j)} = \sqrt{(\mathbf{E}^i \cdot \mathbf{E}^j) A_i A_j} = \sqrt{g^{ij} A_i A_j} = \sqrt{A^j A_j} = \sqrt{A_i A^i} \tag{264}$$

where Eqs. 214 and 228 are used. The magnitude of a vector can also be obtained more directly from the dot product of the covariant and contravariant forms of the vector, that is:

$$|\mathbf{A}| = \sqrt{\mathbf{A} \cdot \mathbf{A}} = \sqrt{(A_i \mathbf{E}^i) \cdot (A^j \mathbf{E}_j)} = \sqrt{(\mathbf{E}^i \cdot \mathbf{E}_j) A_i A^j} = \sqrt{\delta^i_j A_i A^j} = \sqrt{A_i A^i} = \sqrt{A_j A^j} \tag{265}$$

where Eqs. 215 and 172 are used.

4.6.3 Angle between Vectors

The angle θ between two contravariant vectors **A** and **B** is given by:

$$\cos \theta = \frac{\mathbf{A} \cdot \mathbf{B}}{|\mathbf{A}||\mathbf{B}|} = \frac{A^i \mathbf{E}_i \cdot B^j \mathbf{E}_j}{\sqrt{A^k \mathbf{E}_k \cdot A^l \mathbf{E}_l} \sqrt{B^m \mathbf{E}_m \cdot B^n \mathbf{E}_n}} = \frac{g_{ij} A^i B^j}{\sqrt{g_{kl} A^k A^l} \sqrt{g_{mn} B^m B^n}} \tag{266}$$

Similarly, the angle θ between two covariant vectors **A** and **B** is given by:

$$\cos \theta = \frac{\mathbf{A} \cdot \mathbf{B}}{|\mathbf{A}||\mathbf{B}|} = \frac{A_i \mathbf{E}^i \cdot B_j \mathbf{E}^j}{\sqrt{A_k \mathbf{E}^k \cdot A_l \mathbf{E}^l} \sqrt{B_m \mathbf{E}^m \cdot B_n \mathbf{E}^n}} = \frac{g^{ij} A_i B_j}{\sqrt{g^{kl} A_k A_l} \sqrt{g^{mn} B_m B_n}} \tag{267}$$

For two vectors of opposite variance type we have:

$$\cos \theta = \frac{\mathbf{A} \cdot \mathbf{B}}{|\mathbf{A}||\mathbf{B}|} = \frac{A^i B_i}{\sqrt{g_{kl} A^k A^l} \sqrt{g^{mn} B_m B_n}} = \frac{A_i B^i}{\sqrt{g^{kl} A_k A_l} \sqrt{g_{mn} B^m B^n}} \tag{268}$$

All these three cases can be represented by the following formula:

$$\cos \theta = \frac{A^i B_i}{\sqrt{A_l A^l} \sqrt{B_n B^n}} = \frac{A_j B^j}{\sqrt{A_l A^l} \sqrt{B_n B^n}} \tag{269}$$

The angle θ between two sufficiently smooth space curves, C_1 and C_2, intersecting at a given point P in the space is defined as the angle between their tangent vectors, **A** and **B**, at that point (see Figure 17).

4.6.4 Cross Product

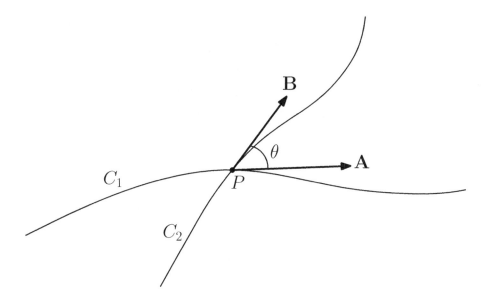

Figure 17: The angle between two space curves, C_1 and C_2, as the angle θ between their tangents, **A** and **B**, at the point of intersection P.

4.6.4 Cross Product

The cross product of two covariant basis vectors in general coordinate systems of a 3D space is given by:

$$\mathbf{E}_i \times \mathbf{E}_j = \frac{\partial x^l}{\partial u^i}\mathbf{e}_l \times \frac{\partial x^m}{\partial u^j}\mathbf{e}_m = \frac{\partial x^l}{\partial u^i}\frac{\partial x^m}{\partial u^j}\mathbf{e}_l \times \mathbf{e}_m = \frac{\partial x^l}{\partial u^i}\frac{\partial x^m}{\partial u^j}\epsilon_{lmn}\mathbf{e}_n \qquad (270)$$

where the indexed x and u are the coordinates of Cartesian and general coordinate systems respectively, the indexed **e** are the Cartesian basis vectors and ϵ_{lmn} is the rank-3 permutation relative tensor as defined by Eq. 155. In the last step of the last equation, Eq. 184 is used to express the cross product of two orthonormal vectors in tensor notation. We note that for orthonormal Cartesian systems, there is no difference between covariant and contravariant tensors and hence $\mathbf{e}_i = \mathbf{e}^i$. We also note that for orthonormal Cartesian systems $g = 1$ where g is the determinant of the covariant metric tensor (see Eqs. 238 and 236).

Now since $\mathbf{e}_n = \mathbf{e}^n = \frac{\partial x^n}{\partial u^k}\mathbf{E}^k$, the last equation becomes:

$$\mathbf{E}_i \times \mathbf{E}_j = \frac{\partial x^l}{\partial u^i}\frac{\partial x^m}{\partial u^j}\frac{\partial x^n}{\partial u^k}\epsilon_{lmn}\mathbf{E}^k = \underline{\epsilon}_{ijk}\mathbf{E}^k \qquad (271)$$

where the underlined absolute covariant permutation tensor is defined as:

$$\underline{\epsilon}_{ijk} = \frac{\partial x^l}{\partial u^i}\frac{\partial x^m}{\partial u^j}\frac{\partial x^n}{\partial u^k}\epsilon_{lmn} \qquad (272)$$

So the final result is:

$$\mathbf{E}_i \times \mathbf{E}_j = \underline{\epsilon}_{ijk}\mathbf{E}^k \qquad (273)$$

4.6.5 Scalar Triple Product

By a similar reasoning, we obtain the following expression for the cross product of two contravariant basis vectors in general coordinate systems:

$$\mathbf{E}^i \times \mathbf{E}^j = \underline{\epsilon}^{ijk}\mathbf{E}_k \tag{274}$$

where the absolute contravariant permutation tensor is defined as:

$$\underline{\epsilon}^{ijk} = \frac{\partial u^i}{\partial x^l}\frac{\partial u^j}{\partial x^m}\frac{\partial u^k}{\partial x^n}\epsilon^{lmn} \tag{275}$$

Considering Eq. 166, the above equations can also be expressed as:

$$\mathbf{E}_i \times \mathbf{E}_j = \underline{\epsilon}_{ijk}\mathbf{E}^k = \sqrt{g}\epsilon_{ijk}\mathbf{E}^k \tag{276}$$

$$\mathbf{E}^i \times \mathbf{E}^j = \underline{\epsilon}^{ijk}\mathbf{E}_k = \frac{\epsilon^{ijk}}{\sqrt{g}}\mathbf{E}_k \tag{277}$$

The cross product of non-basis vectors follows similar rules to those outlined above for the basis vectors; the only difference is that the algebraic product of the components is used as a scale factor for the cross product of their basis vectors. For example, the cross product of two contravariant vectors, A^i and B^j, is given by:

$$\mathbf{A} \times \mathbf{B} = \left(A^i\mathbf{E}_i\right) \times \left(B^j\mathbf{E}_j\right) = A^iB^j\left(\mathbf{E}_i \times \mathbf{E}_j\right) = \underline{\epsilon}_{ijk}A^iB^j\mathbf{E}^k \tag{278}$$

that is, the kth component of this product, which is a vector with covariant components, is:

$$[\mathbf{A} \times \mathbf{B}]_k = \underline{\epsilon}_{ijk}A^iB^j \tag{279}$$

Similarly, the cross product of two covariant vectors, A_i and B_j, is given by:

$$\mathbf{A} \times \mathbf{B} = \left(A_i\mathbf{E}^i\right) \times \left(B_j\mathbf{E}^j\right) = A_iB_j\left(\mathbf{E}^i \times \mathbf{E}^j\right) = \underline{\epsilon}^{ijk}A_iB_j\mathbf{E}_k \tag{280}$$

with the kth contravariant component being given by:

$$[\mathbf{A} \times \mathbf{B}]^k = \underline{\epsilon}^{ijk}A_iB_j \tag{281}$$

4.6.5 Scalar Triple Product

The scalar triple product of three contravariant vectors in a 3D space is given by:

$$\begin{aligned}
\mathbf{A} \cdot (\mathbf{B} \times \mathbf{C}) &= A^i\mathbf{E}_i \cdot \left(B^j\mathbf{E}_j \times C^k\mathbf{E}_k\right) & (282)\\
&= A^iB^jC^k\mathbf{E}_i \cdot \left(\mathbf{E}_j \times \mathbf{E}_k\right) \\
&= A^iB^jC^k\mathbf{E}_i \cdot \left(\underline{\epsilon}_{jkl}\mathbf{E}^l\right) & \text{(Eq. 273)}\\
&= \underline{\epsilon}_{jkl}A^iB^jC^k\left(\mathbf{E}_i \cdot \mathbf{E}^l\right) \\
&= \underline{\epsilon}_{jkl}A^iB^jC^k\delta^l_i & \text{(Eq. 215)}\\
&= \underline{\epsilon}_{jki}A^iB^jC^k & \text{(Eq. 172)}
\end{aligned}$$

4.6.6 Vector Triple Product

$$= \underline{\epsilon}_{ijk} A^i B^j C^k \qquad \text{(Eq. 180)}$$

Similarly, the scalar triple product of three covariant vectors in a 3D space is given by:

$$\begin{aligned}
\mathbf{A} \cdot (\mathbf{B} \times \mathbf{C}) &= A_i \mathbf{E}^i \cdot \left(B_j \mathbf{E}^j \times C_k \mathbf{E}^k \right) & (283) \\
&= A_i B_j C_k \mathbf{E}^i \cdot \left(\mathbf{E}^j \times \mathbf{E}^k \right) & \\
&= A_i B_j C_k \mathbf{E}^i \cdot \left(\underline{\epsilon}^{jkl} \mathbf{E}_l \right) & \text{(Eq. 274)} \\
&= \underline{\epsilon}^{jkl} A_i B_j C_k \left(\mathbf{E}^i \cdot \mathbf{E}_l \right) & \\
&= \underline{\epsilon}^{jkl} A_i B_j C_k \delta^i_l & \text{(Eq. 215)} \\
&= \underline{\epsilon}^{jki} A_i B_j C_k & \text{(Eq. 172)} \\
&= \underline{\epsilon}^{ijk} A_i B_j C_k & \text{(Eq. 180)}
\end{aligned}$$

Other forms can also be obtained by a similar method.

4.6.6 Vector Triple Product

The vector triple product of three vectors, one contravariant and two covariant, in a 3D space is given by:

$$\begin{aligned}
\mathbf{A} \times (\mathbf{B} \times \mathbf{C}) &= A^i \mathbf{E}_i \times \left(B_j \mathbf{E}^j \times C_k \mathbf{E}^k \right) & (284) \\
&= A^i B_j C_k \mathbf{E}_i \times \left(\mathbf{E}^j \times \mathbf{E}^k \right) & \\
&= A^i B_j C_k \mathbf{E}_i \times \left(\underline{\epsilon}^{jkl} \mathbf{E}_l \right) & \text{(Eq. 274)} \\
&= \underline{\epsilon}^{jkl} A^i B_j C_k \left(\mathbf{E}_i \times \mathbf{E}_l \right) & \\
&= \underline{\epsilon}^{jkl} A^i B_j C_k \underline{\epsilon}_{ilm} \mathbf{E}^m & \text{(Eq. 273)} \\
&= \underline{\epsilon}_{ilm} \underline{\epsilon}^{jkl} A^i B_j C_k \mathbf{E}^m & \\
&= \epsilon_{ilm} \epsilon^{jkl} A^i B_j C_k \mathbf{E}^m & \text{(Eq. 166)}
\end{aligned}$$

Following relabeling of indices and writing in a covariant component form, we obtain the following more organized expression:

$$[\mathbf{A} \times (\mathbf{B} \times \mathbf{C})]_i = \epsilon_{ijk} \epsilon^{klm} A^j B_l C_m \qquad (285)$$

We can also obtain a different form from one covariant vector and two contravariant vectors, that is:

$$\begin{aligned}
\mathbf{A} \times (\mathbf{B} \times \mathbf{C}) &= A_i \mathbf{E}^i \times \left(B^j \mathbf{E}_j \times C^k \mathbf{E}_k \right) & (286) \\
&= A_i B^j C^k \mathbf{E}^i \times \left(\mathbf{E}_j \times \mathbf{E}_k \right) & \\
&= A_i B^j C^k \mathbf{E}^i \times \left(\underline{\epsilon}_{jkl} \mathbf{E}^l \right) & \text{(Eq. 273)} \\
&= \underline{\epsilon}_{jkl} A_i B^j C^k \left(\mathbf{E}^i \times \mathbf{E}^l \right) & \\
&= \underline{\epsilon}_{jkl} A_i B^j C^k \underline{\epsilon}^{ilm} \mathbf{E}_m & \text{(Eq. 274)}
\end{aligned}$$

$$= \epsilon^{ilm}\epsilon_{jkl}A_iB^jC^k\mathbf{E}_m$$
$$= \epsilon^{ilm}\epsilon_{jkl}A_iB^jC^k\mathbf{E}_m \qquad (\text{Eq. 166})$$

Following relabeling of indices and writing in a contravariant component form, we obtain the following more organized expression:

$$[\mathbf{A}\times(\mathbf{B}\times\mathbf{C})]^i = \epsilon^{ijk}\epsilon_{klm}A_jB^lC^m \qquad (287)$$

The expressions of the other principal form of the vector triple product, i.e. $(\mathbf{A}\times\mathbf{B})\times\mathbf{C}$, can be obtained by a similar method.

4.6.7 Determinant of Matrix

For a 3×3 matrix representing a rank-2 tensor $\mathbf{A}$ of mixed form in a 3D space, the determinant is given by:

$$\det(\mathbf{A}) = \begin{vmatrix} A_1^1 & A_2^1 & A_3^1 \\ A_1^2 & A_2^2 & A_3^2 \\ A_1^3 & A_2^3 & A_3^3 \end{vmatrix} = \frac{1}{3!}\epsilon^{ijk}\epsilon_{lmn}A_i^lA_j^mA_k^n = \frac{1}{3!}\delta^{ijk}_{lmn}A_i^lA_j^mA_k^n \qquad (288)$$

where Eq. 201 is used in the last step. This may be expressed as a row expansion by substituting $1,2,3$ for l,m,n that is:

$$\det(\mathbf{A}) = \epsilon^{ijk}A_i^1A_j^2A_k^3 \qquad (289)$$

It can also be expressed as a column expansion by substituting $1,2,3$ for i,j,k that is:

$$\det(\mathbf{A}) = \epsilon_{lmn}A_1^lA_2^mA_3^n \qquad (290)$$

More generally, for an $n\times n$ matrix representing a rank-2 tensor $\mathbf{A}$ of mixed form in an nD space, the determinant is given by:

$$\det(\mathbf{A}) = \frac{1}{n!}\epsilon^{i_1\cdots i_n}\epsilon_{j_1\cdots j_n}A_{i_1}^{j_1}\ldots A_{i_n}^{j_n} \qquad (291)$$

Similar definitions can be given using the covariant and contravariant forms of the tensor with the employment of the opposite variance type of the permutation tensors.

4.6.8 Length

The differential of displacement vector in general coordinate systems is given by:

$$d\mathbf{r} = \frac{\partial\mathbf{r}}{\partial u^i}du^i = \mathbf{E}_idu^i = \sum_i |\mathbf{E}_i|\frac{\mathbf{E}_i}{|\mathbf{E}_i|}du^i = \sum_i |\mathbf{E}_i|\hat{\mathbf{E}}_idu^i \qquad (292)$$

where $\mathbf{r}$ is the position vector as defined previously and the hat indicates a normalized vector. The length of line element, ds, which may also be called the differential of arc length, in general coordinate systems is given by:

$$(ds)^2 = d\mathbf{r}\cdot d\mathbf{r} = \mathbf{E}_idu^i\cdot\mathbf{E}_jdu^j = (\mathbf{E}_i\cdot\mathbf{E}_j)du^idu^j = g_{ij}du^idu^j \qquad (293)$$

where g_{ij} is the covariant metric tensor.

For orthogonal coordinate systems, the metric tensor is given by:

$$g_{ij} = \begin{cases} 0 & (i \neq j) \\ (h_i)^2 & (i = j) \end{cases} \qquad (294)$$

where h_i is the scale factor of the ith coordinate. Hence, the last part of Eq. 293 becomes:

$$(ds)^2 = \sum_i (h_i)^2 du^i du^i \qquad (295)$$

with no cross terms (i.e. terms of products involving more than one coordinate like $du^i du^j$ where $i \neq j$) which are generally present in the case of non-orthogonal coordinate systems.

On conducting a transformation from one coordinate system to another coordinate system, where the other system is marked with barred coordinates $\bar{u}$, the length of line element ds will be expressed in the new system as:

$$(ds)^2 = \bar{g}_{ij} d\bar{u}^i d\bar{u}^j \qquad (296)$$

Since the length of line element is an invariant quantity, the same symbol $(ds)^2$ is used in both Eqs. 293 and 296.

Based on the above formulations, the length L of a t-parameterized space curve $C(t)$ defined by $u^i = u^i(t)$ where $i = 1, \cdots, n$, which represents the distance traversed along the curve on moving between its start point P_1 and end point P_2, is given in general coordinate systems of an nD space by:

$$L = \int_C ds = \int_{P_1}^{P_2} \sqrt{g_{ij} du^i du^j} = \int_{t_1}^{t_2} \sqrt{g_{ij} \frac{du^i}{dt} \frac{du^j}{dt}}\, dt \qquad (297)$$

where C represents the space curve, t is a scalar real variable, and t_1 and t_2 are the values of t corresponding to the start and end points respectively. It is noteworthy that some authors add a sign indicator to ensure that the argument of the square root in the above equation is positive. However, as indicated in § 1.1, such a condition is assumed when needed since in this book we deal with real quantities only.

4.6.9 Area

In general coordinate systems of a 3D space, the area of an infinitesimal element of the surface $u^1 = c_1$ in the neighborhood of a given point P, where c_1 is a constant, is obtained by taking the magnitude of the cross product of the differentials of the displacement vectors in the directions of the other two coordinates on that surface at P, i.e. the tangent vectors to the other two coordinate curves at P (see Figure 18). Hence, the area of a differential element on the surface $u^1 = c_1$ is given by:

$$d\sigma(u^1 = c_1) = |\mathbf{dr}_2 \times \mathbf{dr}_3| \qquad (298)$$

4.6.9 Area

$$= \left| \frac{\partial \mathbf{r}}{\partial u^2} \times \frac{\partial \mathbf{r}}{\partial u^3} \right| du^2 du^3$$

$$= |\mathbf{E}_2 \times \mathbf{E}_3| \, du^2 du^3 \qquad \text{(Eq. 45)}$$

$$= \left| \epsilon_{231} \mathbf{E}^1 \right| du^2 du^3 \qquad \text{(Eq. 273)}$$

$$= |\epsilon_{231}| \left| \mathbf{E}^1 \right| du^2 du^3$$

$$= \sqrt{g} \sqrt{\mathbf{E}^1 \cdot \mathbf{E}^1} \, du^2 du^3 \qquad \text{(Eqs. 166 \& 263)}$$

$$= \sqrt{g} \sqrt{g^{11}} \, du^2 du^3 \qquad \text{(Eq. 214)}$$

$$= \sqrt{g g^{11}} \, du^2 du^3$$

where σ represents area and $d\mathbf{r}_2$ and $d\mathbf{r}_3$ are the displacement differentials along the second and third coordinate curves at P while the other symbols are as defined previously.

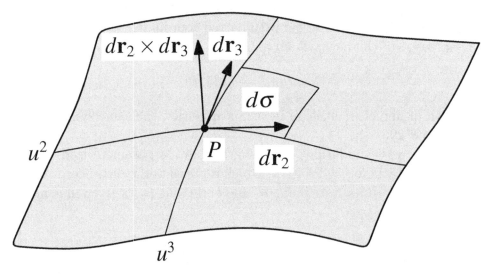

Figure 18: The area $d\sigma$ of an infinitesimal element of the surface $u^1 = c_1$ in the neighborhood of a given point P as the magnitude of the cross product of the differentials of the displacement vectors in the directions of the other two coordinate curves on that surface at P, $d\mathbf{r}_2$ and $d\mathbf{r}_3$.

On generalizing the above argument, the area of a differential element on the surface $u^i = c_i$ ($i = 1, 2, 3$) in a 3D space where c_i is a constant is given by:

$$d\sigma(u^i = c_i) = \sqrt{g g^{ii}} \, du^j du^k \qquad (i \neq j \neq k, \text{ no sum on } i) \qquad (299)$$

In orthogonal coordinate systems in a 3D space we have:

$$\sqrt{g g^{ii}} = \sqrt{(h_i)^2 (h_j)^2 (h_k)^2 \frac{1}{(h_i)^2}} = h_j h_k \qquad (i \neq j \neq k, \text{ no sum on any index}) \qquad (300)$$

where Eqs. 235 and 236 are used. Hence, Eq. 299 becomes:

$$d\sigma(u^i = c_i) = h_j h_k du^j du^k \qquad (i \neq j \neq k, \text{ no sum on any index}) \qquad (301)$$

The last formula represents the area of a surface element of a rectangular shape with sides $h_j du^j$ and $h_k du^k$ (no sum on j or k).

Based on the above discussion, the area A of a finite surface patch can be defined by the following formula:

$$A = \iint_S d\sigma \tag{302}$$

where S represents a surface patch and $d\sigma$ is the area differential of an infinitesimal element on the patch. For coordinate surfaces, the expression for $d\sigma$ can be obtained from the previous formulations.

4.6.10 Volume

In general coordinate systems of a 3D space, the volume of an infinitesimal element of a solid body occupying a given region of the space in the neighborhood of a given point P, where the element is represented by a parallelepiped defined by the three differentials of the displacement vectors $d\mathbf{r}_i$ ($i = 1, 2, 3$) along the three coordinate curves at P, is obtained by taking the magnitude of the scalar triple product of these vectors (refer to § 4.6.5 and see Figure 19). Hence, the volume of a differential element of the body is given by:

$$\begin{aligned}
d\tau &= |d\mathbf{r}_1 \cdot (d\mathbf{r}_2 \times d\mathbf{r}_3)| & (303)\\
&= \left| \frac{\partial \mathbf{r}}{\partial u^1} \cdot \left(\frac{\partial \mathbf{r}}{\partial u^2} \times \frac{\partial \mathbf{r}}{\partial u^3} \right) \right| du^1 du^2 du^3 & \\
&= |\mathbf{E}_1 \cdot (\mathbf{E}_2 \times \mathbf{E}_3)|\, du^1 du^2 du^3 & \text{(Eq. 45)}\\
&= \left| \mathbf{E}_1 \cdot \underline{\epsilon}_{231} \mathbf{E}^1 \right| du^1 du^2 du^3 & \text{(Eq. 273)}\\
&= \left| \mathbf{E}_1 \cdot \mathbf{E}^1 \right| |\underline{\epsilon}_{231}|\, du^1 du^2 du^3 & \\
&= \left| \delta_1^1 \right| |\underline{\epsilon}_{231}|\, du^1 du^2 du^3 & \text{(Eq. 215)}\\
&= \sqrt{g}\, du^1 du^2 du^3 & \text{(Eq. 166)}\\
&= J\, du^1 du^2 du^3 & \text{(Eq. 63)}
\end{aligned}$$

where g is the determinant of the covariant metric tensor g_{ij}, and J is the Jacobian of the transformation as defined previously (see § 2.3). We note that due to the freedom of choice of the order of the variables, which is related to the choice of the coordinate system handedness that could affect the sign of the determinant Jacobian, the sign of the determinant should be adjusted if necessary to have a proper sign for the volume element as a positive quantity. The last line in the last equation is particularly useful for changing the variables in multivariate integrals where the Jacobian facilitates the transformation.

In orthogonal coordinate systems in a 3D space, the above formulation becomes:

$$d\tau = h_1 h_2 h_3\, du^1 du^2 du^3 \tag{304}$$

where h_1, h_2 and h_3 are the scale factors and where Eq. 236 is used. Geometrically, the last formula represents the volume of a rectangular parallelepiped with edges $h_1 du^1$, $h_2 du^2$ and $h_3 du^3$.

4.6.10 Volume

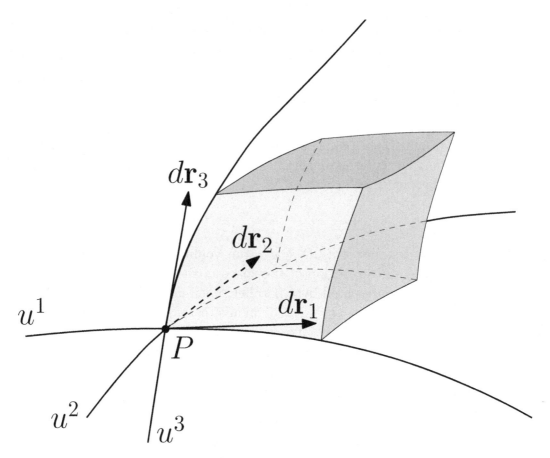

Figure 19: The volume of an infinitesimal element of a solid body in a 3D space in the neighborhood of a given point P as the magnitude of the scalar triple product of the differentials of the displacement vectors in the directions of the three coordinate curves at P, $d\mathbf{r}_1$, $d\mathbf{r}_2$ and $d\mathbf{r}_3$.

The formulae of Eq. 303, which are specific to a 3D space, can be extended to the differential of a generalized volume element in general coordinate systems of an nD space as follows:
$$d\tau = \sqrt{g}\,du^1 \ldots du^n = J\,du^1 \ldots du^n \tag{305}$$
We note that generalized volume elements are used, for instance, to represent the change of variables in multi-variable integrations.

Based on the above discussion, the volume V of a solid body occupying a finite region of the space can be defined as:
$$V = \iiint_\Omega d\tau \tag{306}$$
where Ω represents the region of the space occupied by the solid body and $d\tau$ is the volume differential of an infinitesimal element of the body, and where the expression for $d\tau$ should be obtained from the above formulations.

4.7 Exercises and Revision

4.1 What is special about the Kronecker delta, the permutation and the metric tensors and why they deserve special attention?

4.2 Give detailed definition, in words and in symbols, of the ordinary Kronecker delta tensor in an nD space.

4.3 List and discuss all the main characteristics (e.g. symmetry) of the ordinary Kronecker delta tensor.

4.4 Write the matrix that represents the ordinary Kronecker delta tensor in a 4D space.

4.5 Do we violate the rules of tensor indices when we write: $\delta_{ij} = \delta^{ij} = \delta^i_j = \delta^j_i$?

4.6 Explain the following statement: "The ordinary Kronecker delta tensor is conserved under all proper and improper coordinate transformations". What is the relation between this and the property of isotropy of this tensor?

4.7 List and discuss all the main characteristics (e.g. anti-symmetry) of the permutation tensor.

4.8 What are the other names used to label the permutation tensor?

4.9 Why the rank and the dimension of the permutation tensor are the same? Accordingly, what is the number of components of the rank-2, rank-3 and rank-4 permutation tensors?

4.10 Why the permutation tensor of any rank has only one independent non-vanishing component?

4.11 Prove that the rank-n permutation tensor possesses $n!$ non-zero components.

4.12 Why the permutation tensor is totally anti-symmetric?

4.13 Give the inductive mathematical definition of the components of the permutation tensor of rank-n.

4.14 State the most simple analytical mathematical definition of the components of the permutation tensor of rank-n.

4.15 Make a sketch of the array representing the tank-3 permutation tensor where the nodes of the array are marked with the symbols and values of the components of this tensor.

4.16 Define, mathematically, the rank-n covariant and contravariant absolute permutation tensors, $\underline{\epsilon}_{i_1 \ldots i_n}$ and $\underline{\epsilon}^{i_1 \ldots i_n}$.

4.17 Show that ϵ_{ijk} is a relative tensor of weight -1 and ϵ^{ijk} is a relative tensor of weight $+1$.

4.18 Show that $\underline{\epsilon}_{i_1 \ldots i_n} = \sqrt{g}\, \epsilon_{i_1 \ldots i_n}$ and $\underline{\epsilon}^{i_1 \ldots i_n} = \frac{1}{\sqrt{g}} \epsilon^{i_1 \ldots i_n}$ are absolute tensors assuming that $\epsilon_{i_1 \ldots i_n}$ is a relative tensor of weight -1 and $\epsilon^{i_1 \ldots i_n}$ is a relative tensor of weight $+1$.

4.19 Write $\epsilon^{i_1 i_2 \ldots i_n} \epsilon_{j_1 j_2 \ldots j_n}$ in its determinantal form in terms of the ordinary Kronecker delta.

4.20 Prove the following identity: $\epsilon^{i_1 i_2 \ldots i_n} \epsilon_{i_1 i_2 \ldots i_n} = n!$.

4.21 State a mathematical relation representing the use of the ordinary Kronecker delta tensor acting as an index replacement operator.

4.22 Prove the following relation inductively by writing it in an expanded form in a 3D space: $\delta^i_i = n$.

4.7 Exercises and Revision

4.23 Repeat exercise 4.22 with the relation: $u^i_{,j} = \delta^i_j$ using a matrix form.

4.24 Justify the following relation assuming an orthonormal Cartesian system: $\partial_i x_j = \partial_j x_i$.

4.25 Justify the following relations where the indexed **e** are orthonormal vectors:

$$\mathbf{e}_i \cdot \mathbf{e}_j = \delta_{ij} \qquad\qquad \mathbf{e}_i\mathbf{e}_j : \mathbf{e}_k\mathbf{e}_l = \delta_{ik}\delta_{jl}$$

4.26 Show that $\delta^j_i \delta^k_j \delta^i_k = n$.

4.27 Write the determinantal array form of $\epsilon^{ij}\epsilon_{kl}$ outlining the pattern of the tensor indices in their relation to the indices of the rows and columns of the determinant array.

4.28 Prove the following identity using a truth table: $\epsilon^{ij}\epsilon_{kl} = \delta^i_k \delta^j_l - \delta^i_l \delta^j_k$.

4.29 Prove the following identities justifying each step in your proofs:

$$\epsilon^{il}\epsilon_{kl} = \delta^i_k \qquad\qquad \epsilon^{ijk}\epsilon_{lmk} = \delta^i_l \delta^j_m - \delta^i_m \delta^j_l$$

4.30 Prove the following identities using other more general identities:

$$\epsilon^{ijk}\epsilon_{ljk} = 2\delta^i_l \qquad\qquad \epsilon^{ijk}\epsilon_{ijk} = 6$$

4.31 Outline the similarities and differences between the ordinary Kronecker delta tensor and the generalized Kronecker delta tensor.

4.32 Give the inductive mathematical definition of the generalized Kronecker delta tensor $\delta^{i_1...i_n}_{j_1...j_n}$.

4.33 Write the determinantal array form of the generalized Kronecker delta tensor $\delta^{i_1...i_n}_{j_1...j_n}$ in terms of the ordinary Kronecker delta tensor.

4.34 Define $\epsilon_{i_1...i_n}$ and $\epsilon^{i_1...i_n}$ in terms of the generalized Kronecker delta tensor.

4.35 Prove the relation: $\epsilon^{ijk}\epsilon_{lmn} = \delta^{ijk}_{lmn}$ using an analytic or an inductive or a truth table method.

4.36 Demonstrate that the generalized Kronecker delta is an absolute tensor.

4.37 Prove the following relation justifying each step in your proof: $\delta^{mnq}_{klq} = \delta^{mn}_{kl}$.

4.38 Prove the common form of the epsilon-delta identity.

4.39 Prove the following generalization of the epsilon-delta identity: $g^{ij}\epsilon_{ikl}\epsilon_{jmn} = g_{km}g_{ln} - g_{kn}g_{lm}$.

4.40 List and discuss all the main characteristics (e.g. symmetry) of the metric tensor.

4.41 How many types the metric tensor has?

4.42 Investigate the relation of the metric tensor of a given space to the coordinate systems of the space as well as its relation to the space itself by comparing the characteristics of the metric in different coordinate systems of the space such as being diagonal or not or having constant or variable components and so on. Hence, assess the status of the metric as a property of the space but with a form determined by the adopted coordinate system to describe the space and hence it is also a property of the coordinate system in this sense.

4.43 What is the relation between the covariant metric tensor and the length of an infinitesimal element of arc ds in a general coordinate system?

4.44 How the relation in question 4.43 will become (a) in an orthogonal coordinate system and (b) in an orthonormal Cartesian coordinate system?

4.45 What is the characteristic feature of the metric tensor in orthogonal coordinate systems?
4.46 Write the mathematical expressions for the components of the covariant, contravariant and mixed forms of the metric tensor in terms of the covariant and contravariant basis vectors, $\mathbf{E}_i$ and $\mathbf{E}^i$.
4.47 Write, in full tensor notation, the mathematical expressions for the components of the covariant and contravariant forms of the metric tensor, g_{ij} and g^{ij}.
4.48 What is the relation between the mixed form of the metric tensor and the ordinary Kronecker delta tensor?
4.49 Explain why the metric tensor is not necessarily diagonal in general coordinate systems but it is necessarily symmetric.
4.50 Explain why the diagonal elements of the metric tensor in general coordinate systems are not necessarily of unit magnitude or positive but they are necessarily non-zero.
4.51 Explain why the mixed type metric tensor in any coordinate system is diagonal or in fact it is the unity tensor.
4.52 Show that the covariant and contravariant forms of the metric tensor, g_{ij} and g^{ij}, are inverses of each other.
4.53 Why the determinant of the metric tensor should not vanish at any point in the space?
4.54 If the determinant of the covariant metric tensor g_{ij} is g, what is the determinant of the contravariant metric tensor g^{ij}?
4.55 Show that the metric tensor can be regarded as a transformation of the ordinary Kronecker delta tensor in its different variance types from an orthonormal Cartesian coordinate system to a general coordinate system.
4.56 Justify the use of the metric tensor as an index shifting operator using a mathematical argument.
4.57 Carry out the following index shifting operations recording the order of the indices:

$$g^{ij}C_{klj} \qquad\qquad g_{mn}B^n_{st} \qquad\qquad g^l_n D_{km}{}^n$$

4.58 What is the difference between the three operations in question 4.57?
4.59 Why the order of the raised and lowered indices is important and hence it should be recorded? Mention one form of notation used to record the order of the indices.
4.60 What is the condition that should be satisfied by the metric tensor of a flat space? Give common examples of flat and curved spaces.
4.61 Considering a coordinate transformation, what is the relation between the determinants of the covariant metric tensor in the original and transformed coordinate systems, g and $\bar{g}$?
4.62 $\mathbf{B}$ is a "conjugate" or "associate" tensor of tensor $\mathbf{A}$. What this means?
4.63 Complete and justify the following statement: "The components of the metric tensor are constants *iff* ...etc.".
4.64 What are the covariant and absolute derivatives of the metric tensor?
4.65 Assuming an orthogonal coordinate system of an nD space, complete the following equations where the indexed g represents the metric tensor or its components, $i \neq j$

4.7 Exercises and Revision

in the second equation and there is no sum in the third equation:

$$\det\left(g^{ij}\right)=? \qquad g_{ij}=? \qquad g^{ii}=?$$

4.66 Write, in matrix form, the covariant and contravariant metric tensor for orthonormal Cartesian, cylindrical and spherical coordinate systems. What distinguishes all these matrices? Explain and justify.

4.67 Referring to question 4.66, what is the relation between the diagonal elements of these matrices and the scale factors h_i of the coordinates of these systems?

4.68 Considering the Minkowski metric, is the space of the mechanics of Lorentz transformations flat or curved? Is it homogeneous or not? What effect this can have on the length of element of arc ds?

4.69 Derive the following identities:

$$g_{im}\partial_k g^{mj} = -g^{mj}\partial_k g_{im} \qquad \partial_k g_{ij} = -g_{mj}g_{ni}\partial_k g^{nm}$$

4.70 What is the dot product of **A** and **B** where **A** is a rank-2 covariant tensor and **B** is a contravariant vector? Write this operation in steps providing full justification of each step.

4.71 Derive an expression for the magnitude of a vector **A** when **A** is covariant and when **A** is contravariant.

4.72 Derive an expression for the cosine of the angle θ between two covariant vectors, **A** and **B**, and between two contravariant vectors **C** and **D**.

4.73 What is the meaning of the angle between two intersecting smooth curves?

4.74 What is the cross product of **A** and **B** where these are covariant vectors?

4.75 Complete the following equations assuming a general coordinate system of a 3D space:

$$\mathbf{E}_i \times \mathbf{E}_j =? \qquad \mathbf{E}^i \times \mathbf{E}^j =?$$

4.76 Define the operations of scalar triple product and vector triple product of vectors using tensor language and assuming a general coordinate system of a 3D space.

4.77 What is the relation between the relative and absolute permutation tensors in their covariant and contravariant forms?

4.78 Define the determinant of a matrix **B** in tensor notation assuming a general coordinate system of a 3D space.

4.79 Derive the relation for the length of line element in general coordinate systems: $(ds)^2 = g_{ij}du^i du^j$. How this relation will become when the coordinate system is orthogonal? Justify your answer.

4.80 Write the integral representing the length L of a t-parameterized space curve in terms of the metric tensor.

4.81 Using Eq. 295 plus the scale factors of Table 1, develop expressions for ds in orthonormal Cartesian, cylindrical and spherical coordinate systems.

4.82 Derive the following formula for the area of a differential element on the coordinate surface u^i = constant in a 3D space assuming a general coordinate system:

$$d\sigma(u^i = \text{constant}) = \sqrt{gg^{ii}}\,du^j du^k \qquad (i \neq j \neq k, \text{ no sum on } i)$$

4.7 Exercises and Revision

How this relation will become when the coordinate system is orthogonal?

4.83 Using Eq. 301 plus the scale factors of Table 1, develop expressions for $d\sigma$ on the coordinate surfaces in orthonormal Cartesian, cylindrical and spherical coordinate systems.

4.84 Derive the following formula for the volume of a differential element of a solid body in a 3D space assuming a general coordinate system:

$$d\tau = \sqrt{g}\, du^1 du^2 du^3$$

How this relation will become when the coordinate system is orthogonal?

4.85 Make a plot representing the volume of an infinitesimal element of a solid body in a 3D space as the magnitude of a scalar triple product of three vectors.

4.86 Use the expression of the volume element in general coordinate systems of nD spaces to find the formula for the volume element in orthogonal coordinate systems.

4.87 Using Eq. 304 plus the scale factors of Table 1, develop expressions for $d\tau$ in orthonormal Cartesian, cylindrical and spherical coordinate systems.

Chapter 5
Tensor Differentiation

Ordinary differentiation rules of partial and total derivatives do not satisfy the principle of invariance when they are applied to tensors in general coordinate systems due to the fact that the basis vectors in these systems are coordinate dependent. This means that the ordinary differentiation of non-scalar tensor components in general coordinates does not necessarily results in a tensor. Therefore, special types of differentiation should be defined so that when they apply to tensors they produce tensors. The essence of these operations is to extend the differentiation process to the basis vectors to which the tensor is referred and not only on the tensor components.

The focus of this chapter is the operations of covariant and absolute differentiation which are closely linked. These operations represent generalization of ordinary differentiation in general coordinate systems with an objective of making the derivative of tensors comply with the principle of tensor invariance. Briefly, the differential change of a tensor in general coordinate systems is the result of a change in the basis vectors and a change in the tensor components. Hence, covariant and absolute differentiation, in place of the normal differentiation (i.e. partial and total differentiation respectively), are defined and employed to account for both of these changes by differentiating the basis vectors as well as the components of the tensors.

Since the Christoffel symbols are crucial in the formulation and application of covariant and absolute differentiation, the first section of the present chapter is dedicated to these symbols and their properties. The subsequent two sections will then focus on the covariant differentiation and the absolute differentiation.

5.1 Christoffel Symbols

We start by investigating the main properties of the Christoffel symbols which play crucial roles in tensor calculus in general and are needed for the subsequent development of the upcoming sections since they enter in the definition of covariant and absolute differentiation. The Christoffel symbols are classified as those of the first kind and those of the second kind. These two kinds are linked through the index raising and lowering operators although this does not mean they are general tensors (see next). Both kinds of Christoffel symbols are variable functions of coordinates since they depend in their definition on the metric tensor which is coordinate dependent in general.

The Christoffel symbols of the first and second kind are not general tensors although they are affine tensors of rank-3. We note that affine tensors (see § 3.1.6) are tensors that correspond to admissible linear coordinate transformations from an original rectangular system of coordinates. As a consequence of not being tensors, if all the Christoffel symbols of either kind vanish in a particular coordinate system they will not necessarily vanish in

5.1 Christoffel Symbols

other coordinate systems (see § 3.1.4). For instance, all the Christoffel symbols of both kinds vanish in Cartesian coordinate systems but not in cylindrical or spherical coordinate systems, as will be established later in this section.

The Christoffel symbols of the first kind are defined as:

$$[ij,l] = \frac{1}{2}\left(\partial_j g_{il} + \partial_i g_{jl} - \partial_l g_{ij}\right) \tag{307}$$

where the indexed g is the covariant form of the metric tensor. The Christoffel symbols of the second kind are obtained by raising the third index of the Christoffel symbols of the first kind, and hence they are given by:

$$\Gamma^k_{ij} = g^{kl}\,[ij,l] = \frac{g^{kl}}{2}\left(\partial_j g_{il} + \partial_i g_{jl} - \partial_l g_{ij}\right) \tag{308}$$

where the indexed g is the metric tensor in its contravariant and covariant forms with implied summation over l. Similarly, the Christoffel symbols of the first kind can be obtained from the Christoffel symbols of the second kind by reversing the above process through lowering the upper index, that is:

$$g_{km}\Gamma^k_{ij} = g_{km}g^{kl}\,[ij,l] = \delta^l_m\,[ij,l] = [ij,m] \tag{309}$$

where Eqs. 308, 220 and 172 are used.

The Christoffel symbols of the first and second kind are symmetric in their paired indices, that is:

$$[ij,k] = [ji,k] \tag{310}$$
$$\Gamma^k_{ij} = \Gamma^k_{ji} \tag{311}$$

These properties can be verified by shifting the indices in the definitions of the Christoffel symbols, as given by Eqs. 307 and 308, noting that the metric tensor is symmetric in its two indices (see § 4.5).

For an nD space with n covariant basis vectors $(\mathbf{E}_1, \mathbf{E}_2, \ldots, \mathbf{E}_n)$ spanning the space, the partial derivative $\partial_j \mathbf{E}_i$ for any given i and j is a vector within the space and hence it is in general a linear combination of all the basis vectors. The Christoffel symbols of the second kind are the components of this linear combination, that is:

$$\partial_j \mathbf{E}_i = \Gamma^k_{ij} \mathbf{E}_k \tag{312}$$

Similarly, for the contravariant basis vectors $(\mathbf{E}^1, \mathbf{E}^2, \ldots, \mathbf{E}^n)$ we have:

$$\partial_j \mathbf{E}^i = -\Gamma^i_{kj} \mathbf{E}^k \tag{313}$$

By inner product multiplication of Eq. 312 with $\mathbf{E}^k$ and Eq. 313 with $\mathbf{E}_k$ we obtain:

$$\mathbf{E}^k \cdot \partial_j \mathbf{E}_i = +\Gamma^k_{ij}\mathbf{E}^k \cdot \mathbf{E}_k = +\Gamma^k_{ij}\delta^k_k = +\Gamma^k_{ij} \tag{314}$$

5.1 Christoffel Symbols

$$\mathbf{E}_k \cdot \partial_j \mathbf{E}^i = -\Gamma^i_{kj} \mathbf{E}_k \cdot \mathbf{E}^k = -\Gamma^i_{kj} \delta^k_k = -\Gamma^i_{kj} \tag{315}$$

where Eqs. 49 and 172 are used. These equations reveal that the Christoffel symbols of the second kind are the projections of the partial derivative of the basis vectors in the direction of the basis vectors of the opposite variance type. From Eq. 314 and by using the index lowering operator, we obtain a similar relation that links the Christoffel symbols of the first kind to the basis vectors and their partial derivatives, that is:

$$\mathbf{E}_k \cdot \partial_j \mathbf{E}_i = g_{mk} \mathbf{E}^m \cdot \partial_j \mathbf{E}_i = g_{mk} \Gamma^m_{ij} = [ij, k] \tag{316}$$

where Eq. 309 is used in the last step. This equation reveals that the Christoffel symbols of the first kind are the projections of the partial derivative of the covariant basis vectors in the direction of the basis vectors of the same variance type.

The partial derivative of the components of the covariant metric tensor and the Christoffel symbols of the first kind satisfy the following identity:

$$\partial_k g_{ij} = [ik, j] + [jk, i] \tag{317}$$

This relation can be obtained from the partial derivative of the dot product of the basis vectors with the use of Eq. 316, that is:

$$\begin{aligned}
\partial_k g_{ij} &= \partial_k \left(\mathbf{E}_i \cdot \mathbf{E}_j \right) & \text{(Eq. 213)} \\
&= \left(\partial_k \mathbf{E}_i \right) \cdot \mathbf{E}_j + \mathbf{E}_i \cdot \left(\partial_k \mathbf{E}_j \right) & \text{(product rule)} \\
&= [ik, j] + [jk, i] & \text{(Eq. 316)}
\end{aligned} \tag{318}$$

We note that Eq. 317 is closely linked to the upcoming Ricci theorem (see § 5.2), that is:

$$g_{ij;k} = \partial_k g_{ij} - g_{aj} \Gamma^a_{ik} - g_{ia} \Gamma^a_{jk} = \partial_k g_{ij} - [ik, j] - [jk, i] = 0 \iff \partial_k g_{ij} = [ik, j] - [jk, i] \tag{319}$$

The relation given by Eq. 317 can also be written in terms of the Christoffel symbols of the second kind using the index shifting operator, that is:

$$\partial_k g_{ij} = g_{aj} \Gamma^a_{ik} + g_{ai} \Gamma^a_{jk} \tag{320}$$

where Eq. 309 is used.

Following the method of derivation given in Eq. 318, we obtain the following relation for the partial derivative of the components of the contravariant metric tensor:

$$\begin{aligned}
\partial_k g^{ij} &= \partial_k \left(\mathbf{E}^i \cdot \mathbf{E}^j \right) & \text{(Eq. 214)} \\
&= \left(\partial_k \mathbf{E}^i \right) \cdot \mathbf{E}^j + \mathbf{E}^i \cdot \left(\partial_k \mathbf{E}^j \right) & \text{(product rule)} \\
&= \left(-\Gamma^i_{ak} \mathbf{E}^a \right) \cdot \mathbf{E}^j + \mathbf{E}^i \cdot \left(-\Gamma^j_{ak} \mathbf{E}^a \right) & \text{(Eq. 313)} \\
&= -\Gamma^i_{ak} \mathbf{E}^a \cdot \mathbf{E}^j - \Gamma^j_{ak} \mathbf{E}^i \cdot \mathbf{E}^a \\
&= -g^{aj} \Gamma^i_{ak} - g^{ia} \Gamma^j_{ak} & \text{(Eq. 214)}
\end{aligned} \tag{321}$$

5.1 Christoffel Symbols

that is:
$$\partial_k g^{ij} = -g^{aj}\Gamma^i_{ak} - g^{ia}\Gamma^j_{ak} \tag{322}$$

Like Eq. 317 whose close link to the Ricci theorem is shown in Eq. 319, Eq. 322 can also be seen from this perspective, that is:

$$g^{ij}_{;k} = \partial_k g^{ij} + g^{aj}\Gamma^i_{ak} + g^{ia}\Gamma^j_{ak} = 0 \iff \partial_k g^{ij} = -g^{aj}\Gamma^i_{ak} - g^{ia}\Gamma^j_{ak} \tag{323}$$

The Christoffel symbols of the second kind with two identical indices of opposite variance type satisfy the following relations:

$$\Gamma^j_{ji} = \Gamma^j_{ij} = \frac{1}{2g}\partial_i g = \frac{1}{2}\partial_i (\ln g) = \partial_i (\ln \sqrt{g}) = \frac{1}{\sqrt{g}}\partial_i \sqrt{g} \tag{324}$$

where the main relation can be derived as follows:

$$\Gamma^j_{ij} = \frac{g^{jl}}{2}\left(\partial_j g_{il} + \partial_i g_{jl} - \partial_l g_{ij}\right) \qquad \text{(Eq. 308 with } k=j\text{)} \tag{325}$$

$$= \frac{g^{jl}}{2}\left(\partial_l g_{ij} + \partial_i g_{jl} - \partial_l g_{ij}\right) \qquad \text{(relabeling dummy } j,l \text{ in 1}^{st}\text{ term \& } g^{jl} = g^{lj}\text{)}$$

$$= \frac{1}{2}g^{jl}\partial_i g_{jl}$$

$$= \frac{1}{2g}gg^{jl}\partial_i g_{jl}$$

$$= \frac{1}{2g}\partial_i g \qquad \text{(derivative of determinant)}$$

All the other forms of Eq. 324 can be obtained from the derived form by simple algebraic manipulations with the use of the rules of differentiation and natural logarithms.

In orthogonal coordinate systems, the Christoffel symbols of the first kind can be classified into three main groups, considering the identicality and difference of their indices, which are given by:

$$[ij,i] = [ji,i] = \frac{1}{2}\partial_j g_{ii} \qquad \text{(no sum on } i\text{)} \tag{326}$$

$$[ii,j] = -\frac{1}{2}\partial_j g_{ii} \qquad (i \neq j, \text{ no sum on } i) \tag{327}$$

$$[ij,k] = 0 \qquad (i \neq j \neq k) \tag{328}$$

These three equations can be obtained directly from Eq. 307 with proper labeling of the indices (i.e. according to the labeling of the left hand side of Eqs. 326-328) noting that for orthogonal coordinate systems $g_{ij} = 0$ when $i \neq j$. For example, Eq. 327 can be obtained as follows:

$$[ii,j] = \frac{1}{2}\left(\partial_i g_{ij} + \partial_i g_{ij} - \partial_j g_{ii}\right) \qquad \text{(Eq. 307 with } j \to i \text{ and } l \to j\text{)} \tag{329}$$

$$= \frac{1}{2}\left(0 + 0 - \partial_j g_{ii}\right) \qquad (g_{ij} = 0)$$

5.1 Christoffel Symbols

$$= -\frac{1}{2}\partial_j g_{ii}$$

In fact, Eq. 326 can be obtained from Eq. 317 by exchanging j and k then replacing k with i followed by a simple algebraic manipulation. The middle equality of Eq. 326 is based on the fact that the Christoffel symbols are symmetric in their paired indices (Eq. 310). We note that Eq. 326 includes the case of $j = i$, i.e. when all the three indices are identical, and hence $[ii, i] = \frac{1}{2}\partial_i g_{ii}$. For this reason, we did not add the condition $i \neq j$ to Eq. 326.

Returning to the above indicated consideration of the identicality and difference of the indices of the Christoffel symbols of the first kind, we have 4 main cases: (1) all the indices are identical, (2) only two non-paired indices are identical, (3) only the two paired indices are identical, and (4) all the indices are different. Eq. 326 represents case 1 and case 2, Eq. 327 represents case 3, and Eq. 328 represents case 4. This classification similarly applies to the Christoffel symbols of the second kind as represented by the upcoming Eqs. 331-333.

In orthogonal coordinate systems where $g_{ij} = g^{ij} = 0$ ($i \neq j$), the Christoffel symbols of the second kind are given by:

$$\Gamma^i_{jk} = g^{ii}[jk, i] \qquad \text{(Eq. 308)} \qquad (330)$$
$$= \frac{[jk, i]}{g_{ii}} \qquad \text{(Eq. 235)}$$

with no sum on i. Hence, from Eqs. 326-328 and Eq. 330 the Christoffel symbols of the second kind in orthogonal coordinate systems are given by:

$$\Gamma^i_{ij} = \Gamma^i_{ji} = \frac{g^{ii}}{2}\partial_j g_{ii} = \frac{1}{2g_{ii}}\partial_j g_{ii} = \frac{1}{2}\partial_j \ln g_{ii} \qquad \text{(no sum on } i\text{)} \qquad (331)$$

$$\Gamma^j_{ii} = -\frac{g^{jj}}{2}\partial_j g_{ii} = -\frac{1}{2g_{jj}}\partial_j g_{ii} \qquad (i \neq j, \text{ no sum on } i \text{ or } j) \qquad (332)$$

$$\Gamma^i_{jk} = 0 \qquad (i \neq j \neq k) \qquad (333)$$

where in the last step of Eq. 331 the well known rule of differentiating the natural logarithm is used. Like the Christoffel symbols of the first kind (refer to Eq. 326), Eq. 331 also includes the case of $j = i$, i.e. when all the three indices are identical, that is $\Gamma^i_{ii} = \frac{1}{2g_{ii}}\partial_i g_{ii}$.

In orthogonal coordinate systems of a 3D space, the Christoffel symbols of both kinds vanish when the indices are all different, as seen above (Eqs. 328 and 333). Hence, out of a total of 27 symbols, representing all the possible permutations of the three indices including the repetitive ones, only 21 non-identically vanishing symbols are left since the six non-repetitive permutations are dropped. If we now consider that the Christoffel symbols are symmetric in their paired indices (Eqs. 310 and 311), then we are left with only 15 independent non-identically vanishing symbols since six other permutations representing these symmetric exchanges are also dropped because they are not independent.

5.1 Christoffel Symbols

Accordingly, in orthogonal coordinate systems in a 3D space the 15 independent non-identically vanishing Christoffel symbols of the first kind are:

$$
\begin{aligned}
&[11,1] = +h_1 h_{1,1} &&[11,2] = -h_1 h_{1,2} &&[11,3] = -h_1 h_{1,3} \\
&[12,1] = +h_1 h_{1,2} &&[12,2] = +h_2 h_{2,1} &&[13,1] = +h_1 h_{1,3} \\
&[13,3] = +h_3 h_{3,1} &&[22,1] = -h_2 h_{2,1} &&[22,2] = +h_2 h_{2,2} \\
&[22,3] = -h_2 h_{2,3} &&[23,2] = +h_2 h_{2,3} &&[23,3] = +h_3 h_{3,2} \\
&[33,1] = -h_3 h_{3,1} &&[33,2] = -h_3 h_{3,2} &&[33,3] = +h_3 h_{3,3}
\end{aligned}
\quad (334)
$$

where the indices 1, 2, 3 stand for the three coordinates, h_1, h_2, h_3 are the scale factors corresponding to these coordinates as defined previously (see § 2.5 and 2.6), and the comma indicates, as always, partial derivative with respect to the coordinate represented by the following index. For example, in cylindrical coordinates given by (ρ, ϕ, z), $h_{2,1}$ means the partial derivative of h_2 with respect to the first coordinate and hence $h_{2,1} = \partial_\rho \rho = 1$ since $h_2 = \rho$ (refer to Table 1) and the first coordinate is ρ. Similarly, in spherical coordinates given by (r, θ, ϕ), $h_{3,2}$ means the partial derivative of h_3 with respect to the second coordinate and hence $h_{3,2} = \partial_\theta (r \sin \theta) = r \cos \theta$ since $h_3 = r \sin \theta$ (refer to Table 1) and the second coordinate is θ. As indicated above, because the Christoffel symbols of the first kind are symmetric in their first two indices, the expression of the [21, 1] symbol for instance can be obtained from the expression of the [12, 1] symbol.

The expressions given in Eq. 334 for the Christoffel symbols of the first kind in orthogonal coordinate systems are no more than simple applications of Eqs. 326 and 327 plus Eq. 235. For example, the entry [12, 1] can be obtained as follows:

$$
\begin{aligned}
[12,1] &= \frac{1}{2} \partial_2 g_{11} &&\text{(Eq. 326)} \\
&= \frac{1}{2} \partial_2 (h_1)^2 &&\text{(Eq. 235)} \\
&= h_1 \partial_2 h_1 &&\text{(rules of differentiation)} \\
&= h_1 h_{1,2} &&\text{(notation)}
\end{aligned}
\quad (335)
$$

Similarly, in orthogonal coordinate systems in a 3D space the 15 independent non-identically vanishing Christoffel symbols of the second kind are:

$$
\begin{aligned}
&\Gamma^1_{11} = +\frac{h_{1,1}}{h_1} &&\Gamma^2_{11} = -\frac{h_1 h_{1,2}}{(h_2)^2} &&\Gamma^3_{11} = -\frac{h_1 h_{1,3}}{(h_3)^2} \\
&\Gamma^1_{12} = +\frac{h_{1,2}}{h_1} &&\Gamma^2_{12} = +\frac{h_{2,1}}{h_2} &&\Gamma^1_{13} = +\frac{h_{1,3}}{h_1} \\
&\Gamma^3_{13} = +\frac{h_{3,1}}{h_3} &&\Gamma^1_{22} = -\frac{h_2 h_{2,1}}{(h_1)^2} &&\Gamma^2_{22} = +\frac{h_{2,2}}{h_2} \\
&\Gamma^3_{22} = -\frac{h_2 h_{2,3}}{(h_3)^2} &&\Gamma^2_{23} = +\frac{h_{2,3}}{h_2} &&\Gamma^3_{23} = +\frac{h_{3,2}}{h_3} \\
&\Gamma^1_{33} = -\frac{h_3 h_{3,1}}{(h_1)^2} &&\Gamma^2_{33} = -\frac{h_3 h_{3,2}}{(h_2)^2} &&\Gamma^3_{33} = +\frac{h_{3,3}}{h_3}
\end{aligned}
\quad (336)
$$

5.1 Christoffel Symbols

where the symbols are as explained above. Again, since the Christoffel symbols of the second kind are symmetric in their lower indices, the non-vanishing entries which are not listed above can be obtained from the given entries by permuting the lower indices. The relations of Eq. 336 can be obtained from the relations of Eq. 334 by dividing by $(h_i)^2$ where i is the third index of the Christoffel symbol of the first kind. This can be justified by Eqs. 330 and 235. The relations of Eq. 336 can also be obtained directly from Eqs. 331 and 332 plus Eq. 235, as done for the Christoffel symbols of the first kind where Eqs. 326 and 327 were used.

In any coordinate system, all the Christoffel symbols of the first and second kind vanish identically *iff* all the components of the metric tensor in the given coordinate system are constants. This can be seen from the definitions of the Christoffel symbols, as given by Eqs. 307 and 308, since the partial derivatives will vanish in this case. In affine coordinate systems all the components of the metric tensor are constants and hence all the Christoffel symbols of both kinds vanish identically. The prominent example is the orthonormal Cartesian coordinate systems where all the Christoffel symbols of the first and second kind are identically zero. This can also be seen from Eqs. 334 and 336 since the scale factors are constants (refer to Table 1).

In cylindrical coordinate systems, identified by the coordinates (ρ, ϕ, z), all the Christoffel symbols of the first kind are zero except:

$$[22, 1] = -\rho \tag{337}$$

$$[12, 2] = [21, 2] = \rho \tag{338}$$

where the indices $1, 2$ stand for the coordinates ρ, ϕ respectively. Also, in cylindrical coordinate systems the non-zero Christoffel symbols of the second kind are given by:

$$\Gamma^1_{22} = -\rho \tag{339}$$

$$\Gamma^2_{12} = \Gamma^2_{21} = \frac{1}{\rho} \tag{340}$$

where the symbols are as explained above. Again, these results can be obtained from Eqs. 334 and 336 using the scale factors of Table 1.

In spherical coordinate systems, identified by the coordinates (r, θ, ϕ), the non-zero Christoffel symbols of the first kind are given by:

$$[22, 1] = -r \tag{341}$$

$$[33, 1] = -r \sin^2 \theta \tag{342}$$

$$[12, 2] = [21, 2] = r \tag{343}$$

$$[33, 2] = -r^2 \sin \theta \cos \theta \tag{344}$$

$$[13, 3] = [31, 3] = r \sin^2 \theta \tag{345}$$

$$[23, 3] = [32, 3] = r^2 \sin \theta \cos \theta \tag{346}$$

where the indices $1, 2, 3$ stand for the coordinates r, θ, ϕ respectively. Also, in spherical coordinate systems the non-zero Christoffel symbols of the second kind are given by:

$$\Gamma^1_{22} = -r \tag{347}$$

5.1 Christoffel Symbols

$$\Gamma^1_{33} = -r\sin^2\theta \tag{348}$$

$$\Gamma^2_{12} = \Gamma^2_{21} = \frac{1}{r} \tag{349}$$

$$\Gamma^2_{33} = -\sin\theta\cos\theta \tag{350}$$

$$\Gamma^3_{13} = \Gamma^3_{31} = \frac{1}{r} \tag{351}$$

$$\Gamma^3_{23} = \Gamma^3_{32} = \cot\theta \tag{352}$$

where the symbols are as explained above. As before, these results can be obtained from Eqs. 334 and 336 using the scale factors of Table 1. As seen earlier (refer to § 2.2.2), all these coordinate systems (i.e. orthonormal Cartesian, cylindrical and spherical) are orthogonal systems, and hence Eqs. 334 and 336 do apply.

Because there is an element of arbitrariness in the choice of the order of coordinates, and hence the order of their indices, the Christoffel symbols may be given in terms of coordinate symbols rather than their indices to be more explicit and to avoid ambiguity and confusion. For instance, in the above examples of the cylindrical coordinate systems identified by the coordinates (ρ, ϕ, z) we may use $[\phi\phi, \rho]$ instead of $[22, 1]$ and use $\Gamma^\phi_{\rho\phi}$ instead of Γ^2_{12}. Similarly, for the spherical coordinate systems identified by the coordinates (r, θ, ϕ) we may use $[\theta\theta, r]$ instead of $[22, 1]$ and use $\Gamma^\phi_{r\phi}$ instead of Γ^3_{13}.

The Christoffel symbols of both kinds may also be subscripted or superscripted by the symbol of the metric tensor of the given space (e.g. $[ij,k]_{g_{ij}}$ and $^{g_{ij}}\Gamma^i_{jk}$) to reveal the metric which the symbols are based upon. This is especially important when we have two or more different metrics related to two or more different spaces as it is the case, for instance, in differential geometry of 2D surfaces embedded in a 3D space where we have one metric for the surface and another metric for the 3D space. Alternatively, other means of distinction may be used such as using Latin or upper case indices for the Christoffel symbols of one metric and Greek or lower case indices for the Christoffel symbols of the other metric, e.g. $[ij,k]$ and Γ^i_{jk} for the Christoffel symbols of the space metric and $[\alpha\beta, \gamma]$ and $\Gamma^\gamma_{\alpha\beta}$ for the Christoffel symbols of the surface metric. However, the latter methods do not apply when the indices are numeric rather than symbolic.

The number of independent Christoffel symbols of each kind (first and second) in general coordinate systems is given by:

$$N_{\text{CI}} = \frac{n^2(n+1)}{2} \tag{353}$$

where n is the space dimension. The reason is that, due to the symmetry of the metric tensor there are $\frac{n(n+1)}{2}$ independent metric components, g_{ij}, and for each independent component there are n distinct Christoffel symbols. Alternatively, we have n^3 permutations of the three indices including the repetitive ones, and out of these n^3 permutations we have $(n(n-1)n)$ permutations whose paired indices are different where the three factors correspond to the first, second and third index respectively. Now, due to the symmetry of the Christoffel symbols in their paired indices, half of these $(n(n-1)n)$ permutations are identical to the other half and hence they are not independent. Therefore, the total number of independent Christoffel symbols is:

$$N_{\mathrm{CI}} = n^3 - \frac{(n(n-1)n)}{2} = \frac{2n^3 - n^3 + n^2}{2} = \frac{n^3 + n^2}{2} = \frac{n^2(n+1)}{2} \tag{354}$$

The partial derivative of the Christoffel symbol of the first kind is given by:

$$\partial_k [ij, l] = \frac{1}{2} \left(\partial_k \partial_j g_{il} + \partial_k \partial_i g_{jl} - \partial_k \partial_l g_{ij} \right) \tag{355}$$

where this equation is based on the definition of the Christoffel symbol of the first kind as given by Eq. 307.

Finally, we should remark that there are several notations for the Christoffel symbols. As well as the symbols that we use in this book which may be the most common in use (i.e. $[ij, k]$ for the first kind and Γ_{ij}^k for the second kind), the first kind may also be symbolized as Γ_{ijk} or $\begin{bmatrix} ij \\ k \end{bmatrix}$ while the second kind may be symbolized as $\left\{ {k \atop ij} \right\}$ or $\left\{ {ij \atop k} \right\}$ as well as other notations. There may be some advantages or disadvantages in these different notations. For example, the notations Γ_{ijk} and Γ_{ij}^k may suggest, wrongly, that these symbols are tensors which is not the case since the Christoffel symbols are not general tensors although they are affine tensors. There may also be some advantages in the typesetting and writing of these notations or there are factors related to recognition, readability and even aesthetics.

5.2 Covariant Differentiation

The focus of this section is the operation of covariant differentiation of tensors which is a generalization of the ordinary partial differentiation. The ordinary derivative of a tensor is not a tensor in general. The objective of covariant differentiation is to ensure the invariance of derivative (i.e. being a tensor) in general coordinate systems, and this results in applying more sophisticated rules using Christoffel symbols where different differentiation rules for covariant and contravariant indices apply. The resulting covariant derivative is a tensor which is one rank higher than the differentiated tensor. In brief, the covariant derivative is a partial derivative of the tensor that includes differentiating the basis vectors as well as differentiating the components, as we will see. Hence, the covariant derivative of a general tensor can be given generically by:

$$\partial_k (A_{j,\cdots,n}^{i,\cdots,m} \mathbf{E}_i \mathbf{E}_m \cdots \mathbf{E}^j \mathbf{E}^n) = A_{j,\cdots,n;k}^{i,\cdots,m} \mathbf{E}_i \mathbf{E}_m \cdots \mathbf{E}^j \mathbf{E}^n \tag{356}$$

where the expression of $A_{j,\cdots,n;k}^{i,\cdots,m}$ will be given in the following paragraphs (see e.g. Eq. 366).

More explicitly, the basis vectors in general curvilinear coordinate systems undergo changes in magnitude and direction as they move around in their own space, and hence they are functions of position. These changes should be accounted for when calculating the derivatives of non-scalar tensors in such general systems. Therefore, terms based on using Christoffel symbols are added to the ordinary derivative terms to correct for these changes

5.2 Covariant Differentiation

and this more comprehensive form of derivative is called the covariant derivative. Since in rectilinear coordinate systems the basis vectors are constants, the Christoffel symbol terms vanish identically and hence the covariant derivative reduces to the ordinary partial derivative, but in the other coordinate systems these terms are present in general. As a consequence, the ordinary derivative of a non-scalar tensor is a tensor *iff* the coordinate transformations from Cartesian systems to that system are linear.

It has been suggested that the "covariant" label is an indication that the covariant differentiation operator $\nabla_{;i}$ is in the covariant position. However, it may also be true that "covariant" means "invariant" as the term "covariant" is also used in tensor calculus to mean "invariant". In fact, "covariant" as opposite to "contravariant" applies even to the common form of ordinary partial differentiation since the commonly used partial differential operator ∂_i is also in the covariant position. Anyway, contravariant differentiation, which may be associated with the operator $\nabla^{;j}$, can also be defined for covariant and contravariant tensors by raising the differentiation index of the covariant derivative using the index raising operator, e.g.

$$A_i^{;j} = g^{jk} A_{i;k} \qquad\qquad A^{i;j} = g^{jk} A^i{}_{;k} \qquad (357)$$

However, contravariant differentiation is rarely used.

As an example of how to obtain the covariant derivative of a tensor, let have a vector $\mathbf{A}$ represented by contravariant components in general curvilinear coordinates, that is: $\mathbf{A} = A^i \mathbf{E}_i$. We differentiate this vector following the normal rules of differentiation and taking account of the fact that the basis vectors in general curvilinear coordinate systems are differentiable functions of position and hence they, unlike their rectilinear counterparts, are subject to differentiation using the product rule, that is:

$$\begin{aligned}
\mathbf{A}_{;j} &= \partial_j \mathbf{A} & \text{(definition)} \\
&= \partial_j \left(A^i \mathbf{E}_i \right) \\
&= \mathbf{E}_i \partial_j A^i + A^i \partial_j \mathbf{E}_i & \text{(product rule)} \\
&= \mathbf{E}_i \partial_j A^i + A^i \Gamma^k_{ij} \mathbf{E}_k & \text{(Eq. 312)} \\
&= \mathbf{E}_i \partial_j A^i + A^k \Gamma^i_{kj} \mathbf{E}_i & \text{(relabeling dummy indices } i \leftrightarrow k\text{)} \\
&= \left(\partial_j A^i + A^k \Gamma^i_{kj} \right) \mathbf{E}_i & \text{(taking common factor)} \\
&= A^i{}_{;j} \mathbf{E}_i & \text{(notation)}
\end{aligned} \qquad (358)$$

where $A^i{}_{;j}$, which is a rank-2 mixed tensor, is labeled the "covariant derivative" of A^i.

Similarly, for a vector represented by covariant components in general curvilinear coordinate systems, $\mathbf{A} = A_i \mathbf{E}^i$, we have:

$$\begin{aligned}
\mathbf{A}_{;j} &= \partial_j \mathbf{A} & \text{(definition)} \\
&= \partial_j \left(A_i \mathbf{E}^i \right) \\
&= \mathbf{E}^i \partial_j A_i + A_i \partial_j \mathbf{E}^i & \text{(product rule)} \\
&= \mathbf{E}^i \partial_j A_i - A_i \Gamma^i_{kj} \mathbf{E}^k & \text{(Eq. 313)}
\end{aligned} \qquad (359)$$

5.2 Covariant Differentiation

$$= \mathbf{E}^i \partial_j A_i - A_k \Gamma_{ij}^k \mathbf{E}^i \qquad \text{(relabeling dummy indices } i \leftrightarrow k\text{)}$$
$$= \left(\partial_j A_i - A_k \Gamma_{ij}^k \right) \mathbf{E}^i \qquad \text{(taking common factor)}$$
$$= A_{i;j} \mathbf{E}^i \qquad \text{(notation)}$$

The same rules apply to tensors of higher ranks. For example, for a rank-2 mixed tensor, $\mathbf{A} = A_i{}^j \mathbf{E}^i \mathbf{E}_j$, we have:

$$\mathbf{A}_{;k} = \partial_k \mathbf{A} \qquad \text{(definition)} \qquad (360)$$
$$= \partial_k \left(A_i{}^j \mathbf{E}^i \mathbf{E}_j \right)$$
$$= \left(\partial_k A_i{}^j \right) \mathbf{E}^i \mathbf{E}_j + A_i{}^j \left(\partial_k \mathbf{E}^i \right) \mathbf{E}_j + A_i{}^j \mathbf{E}^i \left(\partial_k \mathbf{E}_j \right) \qquad \text{(product rule)}$$
$$= \left(\partial_k A_i{}^j \right) \mathbf{E}^i \mathbf{E}_j + A_i{}^j \left(-\Gamma_{ak}^i \mathbf{E}^a \right) \mathbf{E}_j + A_i{}^j \mathbf{E}^i \left(\Gamma_{jk}^a \mathbf{E}_a \right) \qquad \text{(Eqs. 312 \& 313)}$$
$$= \left(\partial_k A_i{}^j \right) \mathbf{E}^i \mathbf{E}_j - A_i{}^j \Gamma_{ak}^i \mathbf{E}^a \mathbf{E}_j + A_i{}^j \Gamma_{jk}^a \mathbf{E}^i \mathbf{E}_a \qquad \text{(Eq. 125)}$$
$$= \left(\partial_k A_i{}^j \right) \mathbf{E}^i \mathbf{E}_j - A_a{}^j \Gamma_{ik}^a \mathbf{E}^i \mathbf{E}_j + A_i{}^a \Gamma_{ak}^j \mathbf{E}^i \mathbf{E}_j \qquad \text{(relabeling dummy indices)}$$
$$= \left(\partial_k A_i{}^j - A_a{}^j \Gamma_{ik}^a + A_i{}^a \Gamma_{ak}^j \right) \mathbf{E}^i \mathbf{E}_j \qquad \text{(taking common factor)}$$
$$= A_i{}^j{}_{;k} \mathbf{E}^i \mathbf{E}_j \qquad \text{(notation)}$$

Based on the above arguments and examples, the main rules of covariant differentiation can be outlined in the following examples. For a differentiable vector $\mathbf{A}$, the covariant derivative of the covariant and contravariant forms of the vector is given by:

$$A_{j;i} = \partial_i A_j - \Gamma_{ji}^k A_k \qquad \text{(covariant)} \qquad (361)$$
$$A^j{}_{;i} = \partial_i A^j + \Gamma_{ki}^j A^k \qquad \text{(contravariant)} \qquad (362)$$

Similarly, for a differentiable rank-2 tensor $\mathbf{A}$, the covariant derivative of the covariant, contravariant and mixed forms of the tensor is given by:

$$A_{jk;i} = \partial_i A_{jk} - \Gamma_{ji}^l A_{lk} - \Gamma_{ki}^l A_{jl} \qquad \text{(covariant)} \qquad (363)$$
$$A^{jk}{}_{;i} = \partial_i A^{jk} + \Gamma_{li}^j A^{lk} + \Gamma_{li}^k A^{jl} \qquad \text{(contravariant)} \qquad (364)$$
$$A_j{}^k{}_{;i} = \partial_i A_j{}^k + \Gamma_{li}^k A_j{}^l - \Gamma_{ji}^l A_l{}^k \qquad \text{(mixed)} \qquad (365)$$

We note that for the mixed form there are two possibilities: one associated with the dyad $\mathbf{E}^j \mathbf{E}_k$ and the other with the dyad $\mathbf{E}_k \mathbf{E}^j$.

Following the methods and techniques outlined in the previous examples, we can easily deduce the pattern of the operation of covariant differentiation. To obtain the covariant derivative of a tensor in general, we start with an ordinary partial derivative term of the component of the given tensor. Then for each tensor index an extra Christoffel symbol term is added, positive for contravariant indices and negative for covariant indices, where the differentiation index is one of the lower indices in the Christoffel symbol of the second kind. Hence, for a differentiable rank-n tensor $\mathbf{A}$ in general coordinate systems the covariant derivative with respect to the qth coordinate is given by:

$$A^{ij\ldots k}_{lm\ldots p;q} = \partial_q A^{ij\ldots k}_{lm\ldots p} + \Gamma_{aq}^i A^{aj\ldots k}_{lm\ldots p} + \Gamma_{aq}^j A^{ia\ldots k}_{lm\ldots p} + \cdots + \Gamma_{aq}^k A^{ij\ldots a}_{lm\ldots p} \qquad (366)$$

5.2 Covariant Differentiation

$$-\Gamma^a_{lq}A^{ij...k}_{am...p} - \Gamma^a_{mq}A^{ij...k}_{la...p} - \cdots - \Gamma^a_{pq}A^{ij...k}_{lm...a}$$

In fact, there is practically only one possibility for the arrangement and labeling of the indices in the Christoffel symbol terms of the covariant derivative of a tensor of any rank if the following rules are observed:

1. The second subscript index of the Christoffel symbol is the differentiation index.
2. The differentiated index of the tensor in the Christoffel symbol term is contracted with one of the indices of the Christoffel symbol using a new label and hence they are opposite in their covariant and contravariant type.
3. The label of the differentiated index is transferred from the tensor to the Christoffel symbol keeping its position as covariant or contravariant.
4. All the other indices of the tensor in the concerned Christoffel symbol term keep their labels, position and order.

The ordinary partial derivative term in the covariant derivative expressions (see e.g. Eq. 366) represents the rate of change of the tensor components with change of position as a result of moving along the coordinate curve of the differentiated index, while the Christoffel symbol terms represent the change experienced by the local basis vectors as a result of the same movement. This can be seen from the development of Eqs. 358-360 where the indicated terms correspond to the components and basis vectors according to the product rule of differentiation.

From the above discussion, it is obvious that to obtain the covariant derivative, the Christoffel symbols of the second kind should be obtained and these symbols are dependent on the metric tensor. Hence, the covariant derivative is dependent on having the space metric corresponding to the particular coordinate system. We also note that the covariant derivative of a tensor is a tensor whose covariant rank is higher than the covariant rank of the original tensor by one. Hence, the covariant derivative of a rank-n tensor of type (r,s) is a rank-$(n+1)$ tensor of type $(r,s+1)$.

In all coordinate systems, the covariant derivative of a differentiable scalar function of position, f, is the same as the ordinary partial derivative, that is:

$$f_{;i} = f_{,i} = \partial_i f \tag{367}$$

This is justified by the fact that the covariant derivative is different from the ordinary partial derivative because the basis vectors in general curvilinear coordinate systems are dependent on their spatial position, and since a scalar is independent of the basis vectors the covariant derivative and the partial derivative are identical. The derivation of the expressions of the covariant derivative of contravariant and covariant vectors and higher rank tensors, as given by Eqs. 358-360, clearly justifies this logic where the product rule does not apply due to the absence of a basis vector in the representation of a scalar. This can also be concluded from the covariant derivative rules as demonstrated in the previous statements and formulated in the above equations like Eq. 366 since a scalar has no free index and hence it cannot have any Christoffel symbol term. By a similar reasoning, since the Christoffel symbols are identically zero in rectilinear coordinate systems, the covariant derivative in these systems is the same as the ordinary partial derivative for all tensor

5.2 Covariant Differentiation

ranks, whether scalars or not. This can also be seen from the product rule (as employed in the development of Eqs. 358-360 for instance) where all the terms involving differentiation of basis vectors will vanish since these vectors are constant in rectilinear systems.

Another important fact about covariant differentiation is that the covariant derivative of the metric tensor in its covariant, contravariant and mixed forms is zero in all coordinate systems and hence it is treated like a constant with respect to covariant differentiation. Accordingly, the covariant derivative operator bypasses the metric tensor, e.g.

$$\partial_{;m} \left(g_{ij} A^j \right) = g_{ij} \partial_{;m} A^j \tag{368}$$

and hence the metric tensor commutes with the covariant differential operator. We will expand on this issue later in this section.

Several rules of ordinary differentiation are naturally extended to covariant differentiation. For example, covariant differentiation is a linear operation with respect to algebraic sums of tensor terms and hence the covariant derivative of a sum is the sum of the covariant derivatives of the terms, that is:[22]

$$(a\mathbf{A} \pm b\mathbf{B})_{;i} = a\mathbf{A}_{;i} \pm b\mathbf{B}_{;i} \tag{369}$$

where a and b are scalar constants and $\mathbf{A}$ and $\mathbf{B}$ are differentiable tensors. The product rule of ordinary differentiation also applies to the covariant differentiation of inner and outer products of tensors, that is:

$$(\mathbf{A} \circ \mathbf{B})_{;i} = \mathbf{A}_{;i} \circ \mathbf{B} + \mathbf{A} \circ \mathbf{B}_{;i} \tag{370}$$

where the symbol ∘ denotes an inner or outer product operator. However, as seen in this equation, the order of the tensors should be observed since tensor multiplication, unlike ordinary algebraic multiplication, is not commutative (refer to § 3.2.3). The product rule is valid for the inner product of tensors because the inner product is an outer product operation followed by a contraction of indices, and covariant differentiation and contraction of indices do commute as we will see later.

A principal difference between partial differentiation and covariant differentiation is that for successive differential operations with respect to different indices the ordinary partial derivative operators do commute with each other, assuming that the well known continuity condition is satisfied, but the covariant differential operators do not commute, that is:

$$\partial_i \partial_j = \partial_j \partial_i \qquad \partial_{;i} \partial_{;j} \neq \partial_{;j} \partial_{;i} \tag{371}$$

This will be verified later in this section.

As indicated earlier, according to the "Ricci theorem" the covariant derivative of the covariant, contravariant and mixed forms of the metric tensor is zero. This has nothing to do with the metric tensor being a constant function of coordinates, which is true only for

---

[22] We use a semicolon with symbolic notation of tensors in this equation and other similar equations for the sake of clarity; the meaning should be obvious.

5.2 Covariant Differentiation

rectilinear coordinate systems, but this arises from the fact that the covariant derivative quantifies the change with position of the basis vectors in magnitude and direction as well as the change in components, and these contributions in the case of the metric tensor cancel each other resulting in a total null effect. For example, for the covariant form of the metric tensor we have:

$$
\begin{aligned}
\mathbf{g}_{;k} &= \partial_k \left(g_{ij} \mathbf{E}^i \mathbf{E}^j \right) && \text{(definition)} \\
&= (\partial_k g_{ij}) \mathbf{E}^i \mathbf{E}^j + g_{ij} \left(\partial_k \mathbf{E}^i \right) \mathbf{E}^j + g_{ij} \mathbf{E}^i \left(\partial_k \mathbf{E}^j \right) && \text{(product rule)} \\
&= ([ik,j] + [jk,i]) \mathbf{E}^i \mathbf{E}^j + g_{ij} \left(-\Gamma^i_{lk} \mathbf{E}^l \right) \mathbf{E}^j + g_{ij} \mathbf{E}^i \left(-\Gamma^j_{lk} \mathbf{E}^l \right) && \text{(Eqs. 317 \& 313)} \\
&= [ik,j] \mathbf{E}^i \mathbf{E}^j + [jk,i] \mathbf{E}^i \mathbf{E}^j - g_{ij} \Gamma^i_{lk} \mathbf{E}^l \mathbf{E}^j - g_{ij} \Gamma^j_{lk} \mathbf{E}^i \mathbf{E}^l \\
&= [ik,j] \mathbf{E}^i \mathbf{E}^j + [jk,i] \mathbf{E}^i \mathbf{E}^j - [lk,j] \mathbf{E}^l \mathbf{E}^j - [lk,i] \mathbf{E}^i \mathbf{E}^l && \text{(Eq. 309)} \\
&= [ik,j] \mathbf{E}^i \mathbf{E}^j + [jk,i] \mathbf{E}^i \mathbf{E}^j - [ik,j] \mathbf{E}^i \mathbf{E}^j - [jk,i] \mathbf{E}^i \mathbf{E}^j && \text{(relabeling dummy } l\text{)} \\
&= \mathbf{0}
\end{aligned}
\tag{372}
$$

Similarly, for the contravariant form of the metric tensor we have:

$$
\begin{aligned}
\mathbf{g}_{;k} &= \partial_k \left(g^{ij} \mathbf{E}_i \mathbf{E}_j \right) && \text{(definition)} \\
&= \left(\partial_k g^{ij} \right) \mathbf{E}_i \mathbf{E}_j + g^{ij} \left(\partial_k \mathbf{E}_i \right) \mathbf{E}_j + g^{ij} \mathbf{E}_i \left(\partial_k \mathbf{E}_j \right) && \text{(product rule)} \\
&= \left(-g^{aj} \Gamma^i_{ak} - g^{ia} \Gamma^j_{ak} \right) \mathbf{E}_i \mathbf{E}_j + g^{ij} \left(\Gamma^a_{ik} \mathbf{E}_a \right) \mathbf{E}_j + g^{ij} \mathbf{E}_i \left(\Gamma^a_{jk} \mathbf{E}_a \right) && \text{(Eqs. 322 \& 312)} \\
&= -g^{aj} \Gamma^i_{ak} \mathbf{E}_i \mathbf{E}_j - g^{ia} \Gamma^j_{ak} \mathbf{E}_i \mathbf{E}_j + g^{ij} \Gamma^a_{ik} \mathbf{E}_a \mathbf{E}_j + g^{ij} \Gamma^a_{jk} \mathbf{E}_i \mathbf{E}_a \\
&= -g^{aj} \Gamma^i_{ak} \mathbf{E}_i \mathbf{E}_j - g^{ia} \Gamma^j_{ak} \mathbf{E}_i \mathbf{E}_j + g^{aj} \Gamma^i_{ak} \mathbf{E}_i \mathbf{E}_j + g^{ia} \Gamma^j_{ak} \mathbf{E}_i \mathbf{E}_j && \text{(relabeling indices)} \\
&= \mathbf{0}
\end{aligned}
\tag{373}
$$

As for the mixed form of the metric tensor we have:

$$
\begin{aligned}
\mathbf{g}_{;k} &= \partial_k \left(\delta^i{}_j \mathbf{E}_i \mathbf{E}^j \right) && \text{(definition)} \\
&= \left(\partial_k \delta^i{}_j \right) \mathbf{E}_i \mathbf{E}^j + \delta^i{}_j \left(\partial_k \mathbf{E}_i \right) \mathbf{E}^j + \delta^i{}_j \mathbf{E}_i \left(\partial_k \mathbf{E}^j \right) && \text{(product rule)} \\
&= \mathbf{0} + \delta^i{}_j \left(\Gamma^a_{ik} \mathbf{E}_a \right) \mathbf{E}^j + \delta^i{}_j \mathbf{E}_i \left(-\Gamma^j_{ak} \mathbf{E}^a \right) && \text{(Eqs. 312 \& 313)} \\
&= \delta^i{}_j \Gamma^a_{ik} \mathbf{E}_a \mathbf{E}^j - \delta^i{}_j \Gamma^j_{ak} \mathbf{E}_i \mathbf{E}^a \\
&= \Gamma^a_{jk} \mathbf{E}_a \mathbf{E}^j - \Gamma^i_{ak} \mathbf{E}_i \mathbf{E}^a && \text{(Eq. 172)} \\
&= \Gamma^i_{jk} \mathbf{E}_i \mathbf{E}^j - \Gamma^i_{jk} \mathbf{E}_i \mathbf{E}^j && \text{(relabeling dummy indices)} \\
&= \mathbf{0}
\end{aligned}
\tag{374}
$$

As a result of the Ricci theorem, the metric tensor behaves as a constant with respect to the covariant derivative operation, that is:

$$
\mathbf{g}_{;k} = \mathbf{0}
\tag{375}
$$

where $\mathbf{g}$ is the metric tensor in its covariant or contravariant or mixed form, as seen above. Consequently, the covariant derivative operator bypasses the metric tensor, that is:

$$
(\mathbf{g} \circ \mathbf{A})_{;k} = \mathbf{g} \circ \mathbf{A}_{;k}
\tag{376}
$$

5.2 Covariant Differentiation

where **A** is a general tensor and the symbol ○ denotes an inner or outer tensor product.[23] The commutativity of the covariant derivative operator and the metric tensor acting as an index shifting operator may be demonstrated more vividly by using the indicial notation, e.g.

$$g_{ik}\left(A^k_{\;;j}\right) = g_{ik}A^k_{\;;j} = A_{i;j} = (A_i)_{;j} = \left(g_{ik}A^k\right)_{;j} \tag{377}$$

$$g_{mi}g_{nj}\left(A^{mn}_{\;\;;k}\right) = g_{mi}g_{nj}A^{mn}_{\;\;;k} = A_{ij;k} = (A_{ij})_{;k} = (g_{mi}g_{nj}A^{mn})_{;k} \tag{378}$$

where we see that the sequence of these operations, seen in one order from the right and in another order from the left, has no effect on the final result in the middle.

Like the metric tensor, the ordinary Kronecker delta tensor is constant with regard to covariant differentiation and hence the covariant derivative of the Kronecker delta is identically zero, that is:[24]

$$\begin{aligned}
\delta^i_{j;k} &= \partial_k \delta^i_j + \delta^a_j \Gamma^i_{ak} - \delta^i_a \Gamma^a_{jk} &&\text{(Eq. 365)} \\
&= 0 + \delta^a_j \Gamma^i_{ak} - \delta^i_a \Gamma^a_{jk} &&(\delta^i_j \text{ is constant}) \\
&= 0 + \Gamma^i_{jk} - \Gamma^i_{jk} &&\text{(Eq. 172)} \\
&= 0
\end{aligned} \tag{379}$$

Accordingly, the covariant differential operator bypasses the Kronecker delta tensor which is involved in inner and outer tensor products:[25]

$$(\boldsymbol{\delta} \circ \mathbf{A})_{;k} = \boldsymbol{\delta} \circ \mathbf{A}_{;k} \tag{380}$$

The rule of the Kronecker delta may be regarded as an instance of the rule of the metric tensor, as stated by the Ricci theorem, since the Kronecker delta is a metric tensor for certain systems and types. Like the ordinary Kronecker delta, the covariant derivative of the generalized Kronecker delta is also identically zero. This may be deduced from Eq. 379 plus Eq. 199 where the generalized Kronecker delta is given as a determinant consisting of ordinary Kronecker deltas and hence it is a sum of products of ordinary Kronecker deltas whose partial derivative vanishes because each term in the derivative contains a derivative of an ordinary Kronecker delta.

Based on the previous statements, we conclude that covariant differentiation and contraction of index operations commute with each other, e.g.

$$\left(A^{ij}_{k;l}\right)\delta^k_j = A^{ij}_{k;l}\delta^k_j = A^{ik}_{k;l} = \left(A^{ik}_k\right)_{;l} = \left(A^{ij}_k \delta^k_j\right)_{;l} \tag{381}$$

where we see again that the different sequences from the right and from the left produce the same result in the middle. For clarity, we represented the contraction operation in this

---

[23] Although the metric tensor is normally used in inner product operations for raising and lowering indices, the possibility of its involvement in outer product operations should not be ruled out.

[24] For diversity, we use indicial notation rather than symbolic notation which we used for example in verifying $\mathbf{g}_{;k} = \mathbf{0}$.

[25] Like the metric tensor, the Kronecker delta tensor is normally used in inner product operations for replacement of indices; however the possibility of its involvement in outer product operations should not be ruled out.

5.2 Covariant Differentiation

example by an inner product operation of the ordinary Kronecker delta tensor with the contracted tensor.

For a differentiable function $f(x, y)$ of class C^2 (i.e. all the second order partial derivatives of the function do exist and are continuous), the mixed partial derivatives are equal, that is:

$$\partial_x \partial_y f = \partial_y \partial_x f \tag{382}$$

However, even if the components of a tensor satisfy this condition (i.e. being of class C^2), this is not sufficient for the equality of the mixed covariant derivatives. What is required for the mixed covariant derivatives to be equal is the vanishing of the Riemann-Christoffel curvature tensor (see § 7.2.1) which is equivalent to having an intrinsically flat space. This will be verified later in this section.

Higher order covariant derivatives are defined as derivatives of derivatives by successive repetition of the process of covariant differentiation. However, the order of differentiation, in the case of differentiating with respect to different indices, should be respected as stated in the previous statements. For example, the second order mixed jk covariant derivative of a contravariant vector $\mathbf{A}$ is given by:

$$\begin{aligned} A^i_{;jk} &= \left(A^i_{;j}\right)_{;k} \\ &= \partial_k A^i_{;j} + \Gamma^i_{ak} A^a_{;j} - \Gamma^a_{jk} A^i_{;a} \\ &= \partial_k \left(\partial_j A^i + \Gamma^i_{aj} A^a\right) + \Gamma^i_{ak} \left(\partial_j A^a + \Gamma^a_{bj} A^b\right) - \Gamma^a_{jk} \left(\partial_a A^i + \Gamma^i_{ba} A^b\right) \\ &= \partial_k \partial_j A^i + \Gamma^i_{aj} \partial_k A^a + A^a \partial_k \Gamma^i_{aj} + \Gamma^i_{ak} \partial_j A^a + \Gamma^i_{ak} \Gamma^a_{bj} A^b - \Gamma^a_{jk} \partial_a A^i - \Gamma^a_{jk} \Gamma^i_{ba} A^b \end{aligned} \tag{383}$$

that is:

$$A^i_{;jk} = \partial_k \partial_j A^i + \Gamma^i_{aj} \partial_k A^a - \Gamma^a_{jk} \partial_a A^i + \Gamma^i_{ak} \partial_j A^a + A^a \left(\partial_k \Gamma^i_{aj} - \Gamma^b_{jk} \Gamma^i_{ba} + \Gamma^i_{bk} \Gamma^b_{aj}\right) \tag{384}$$

The second order mixed kj covariant derivative of a contravariant vector can be derived similarly. However, it can be obtained more easily from the last equation by interchanging the j and k indices, that is:

$$A^i_{;kj} = \partial_j \partial_k A^i + \Gamma^i_{ak} \partial_j A^a - \Gamma^a_{kj} \partial_a A^i + \Gamma^i_{aj} \partial_k A^a + A^a \left(\partial_j \Gamma^i_{ak} - \Gamma^b_{kj} \Gamma^i_{ba} + \Gamma^i_{bj} \Gamma^b_{ak}\right) \tag{385}$$

The inequality of the jk and kj mixed derivatives in general can be verified by subtracting the two sides of the last two equations from each other where the right hand side will not vanish.

Similarly, the second order mixed jk covariant derivative of a covariant vector $\mathbf{A}$ is given by:

$$\begin{aligned} A_{i;jk} &= \left(A_{i;j}\right)_{;k} \\ &= \partial_k A_{i;j} - \Gamma^a_{ik} A_{a;j} - \Gamma^a_{jk} A_{i;a} \\ &= \partial_k \left(\partial_j A_i - \Gamma^b_{ij} A_b\right) - \Gamma^a_{ik} \left(\partial_j A_a - \Gamma^b_{aj} A_b\right) - \Gamma^a_{jk} \left(\partial_a A_i - \Gamma^b_{ia} A_b\right) \\ &= \partial_k \partial_j A_i - \Gamma^b_{ij} \partial_k A_b - A_b \partial_k \Gamma^b_{ij} - \Gamma^a_{ik} \partial_j A_a + \Gamma^a_{ik} \Gamma^b_{aj} A_b - \Gamma^a_{jk} \partial_a A_i + \Gamma^a_{jk} \Gamma^b_{ia} A_b \end{aligned} \tag{386}$$

5.2 Covariant Differentiation

that is:

$$A_{i;jk} = \partial_k\partial_j A_i - \Gamma^a_{ij}\partial_k A_a - \Gamma^a_{ik}\partial_j A_a - \Gamma^a_{jk}\partial_a A_i - A_a\left(\partial_k\Gamma^a_{ij} - \Gamma^b_{ik}\Gamma^a_{bj} - \Gamma^b_{jk}\Gamma^a_{ib}\right) \quad (387)$$

The second order mixed kj covariant derivative of a covariant vector can be obtained from the last equation by interchanging the j and k indices, that is:

$$A_{i;kj} = \partial_j\partial_k A_i - \Gamma^a_{ik}\partial_j A_a - \Gamma^a_{ij}\partial_k A_a - \Gamma^a_{kj}\partial_a A_i - A_a\left(\partial_j\Gamma^a_{ik} - \Gamma^b_{ij}\Gamma^a_{bk} - \Gamma^b_{kj}\Gamma^a_{ib}\right) \quad (388)$$

Again, the inequality of the jk and kj mixed derivatives can be verified by subtracting the two sides of the last two equations from each other where the right hand side will not vanish.

We will see in § 7.2.1 that the mixed second order covariant derivatives of a covariant vector A_i are linked through the Riemann-Christoffel curvature tensor R^a_{ijk} by the following relation:

$$A_{i;jk} - A_{i;kj} = A_a R^a_{ijk} \quad (389)$$

This relation can be verified by subtracting the two sides of Eq. 388 from the two sides of Eq. 387 and employing the definition of the Riemann-Christoffel curvature tensor of Eq. 560 to the right hand side, that is:

$$\begin{aligned} A_{i;jk} - A_{i;kj} &= -A_a\left(\partial_k\Gamma^a_{ij} - \Gamma^b_{ik}\Gamma^a_{bj} - \Gamma^b_{jk}\Gamma^a_{ib}\right) + A_a\left(\partial_j\Gamma^a_{ik} - \Gamma^b_{ij}\Gamma^a_{bk} - \Gamma^b_{kj}\Gamma^a_{ib}\right) \quad (390) \\ &= A_a\left(\partial_j\Gamma^a_{ik} - \Gamma^b_{ij}\Gamma^a_{bk} - \Gamma^b_{kj}\Gamma^a_{ib}\right) - A_a\left(\partial_k\Gamma^a_{ij} - \Gamma^b_{ik}\Gamma^a_{bj} - \Gamma^b_{jk}\Gamma^a_{ib}\right) \\ &= A_a\left(\partial_j\Gamma^a_{ik} - \Gamma^b_{ij}\Gamma^a_{bk} - \Gamma^b_{kj}\Gamma^a_{ib} - \partial_k\Gamma^a_{ij} + \Gamma^b_{ik}\Gamma^a_{bj} + \Gamma^b_{jk}\Gamma^a_{ib}\right) \\ &= A_a\left(\partial_j\Gamma^a_{ik} - \Gamma^b_{ij}\Gamma^a_{bk} - \partial_k\Gamma^a_{ij} + \Gamma^b_{ik}\Gamma^a_{bj}\right) \\ &= A_a R^a_{ijk} \end{aligned}$$

where the last step is based on Eq. 560.

The covariant derivatives of relative tensors, which are also relative tensors of the same weight as the original tensors, are obtained by adding a weight term to the normal formulae of covariant derivative. Hence, the covariant derivative of a relative scalar with weight w is given by:

$$f_{;i} = f_{,i} - wf\Gamma^j_{ji} \quad (391)$$

while the covariant derivative of relative tensors of higher ranks with weight w is obtained by adding the following term to the right hand side of Eq. 366:

$$-wA^{ij...k}_{lm...p}\Gamma^a_{aq} \quad (392)$$

Unlike ordinary differentiation, the covariant derivative of a non-scalar tensor with constant components is not zero in general curvilinear coordinate systems due to the presence of the Christoffel symbols in the definition of the covariant derivative, as given by Eq. 366. More explicitly, even though the partial derivative term is zero because the components are constant, the Christoffel symbol terms are not zero in general because the basis vectors are variables (refer to the derivation in Eqs. 358-360).

5.3 Absolute Differentiation   129

In rectilinear coordinate systems, the Christoffel symbols are identically zero because the basis vectors are constants, and hence the covariant derivative is the same as the ordinary partial derivative for all tensor ranks. As a result, when the components of the metric tensor g_{ij} are constants, as it is the case for example in rectangular coordinate systems, the covariant derivative becomes ordinary partial derivative. The constancy of the metric tensor in this case can be concluded from the definition of the components of the metric tensor as dot products of the basis vectors, as seen for example in Eqs. 213 and 214 (also see Table 1).

We remark that for a differentiable covariant vector **A** which is a gradient of a scalar field we have:
$$A_{i;j} = A_{j;i} \tag{393}$$
This can be easily verified by defining **A** as: $A_i = f_{,i}$ where f is a scalar and hence we have:

$$\begin{aligned}
A_{i;j} &= (f_{,i})_{;j} & & (394)\\
&= (f_{,i})_{,j} - \Gamma_{ij}^k f_{,k} & & \text{(Eq. 361)}\\
&= f_{,ij} - \Gamma_{ij}^k f_{,k} & &\\
&= f_{,ji} - \Gamma_{ji}^k f_{,k} & & \text{(Eqs. 382 \& 311)}\\
&= (f_{,j})_{;i} & & \text{(Eq. 361)}\\
&= A_{j;i} & &
\end{aligned}$$

Another important remark is that the covariant derivative of the basis vectors of the covariant and contravariant types is identically zero, that is:

$$\mathbf{E}_{i;j} = \partial_j \mathbf{E}_i - \Gamma_{ij}^k \mathbf{E}_k = +\Gamma_{ij}^k \mathbf{E}_k - \Gamma_{ij}^k \mathbf{E}_k = \mathbf{0} \tag{395}$$
$$\mathbf{E}^i_{;j} = \partial_j \mathbf{E}^i + \Gamma_{kj}^i \mathbf{E}^k = -\Gamma_{kj}^i \mathbf{E}^k + \Gamma_{kj}^i \mathbf{E}^k = \mathbf{0} \tag{396}$$

where Eqs. 361 and 362 are used in the first steps, while Eqs. 312 and 313 are used in the second steps.

5.3 Absolute Differentiation

The absolute derivative of a tensor along a t-parameterized curve $C(t)$ in an nD space with respect to the parameter t is the inner product of the covariant derivative of the tensor and the tangent vector to the curve. In brief, the absolute derivative is a covariant derivative of a tensor along a curve. For a tensor A^i, the inner product of $A^i_{;j}$, which is a tensor, with another tensor is a tensor. Now, if the other tensor is $\frac{du^i}{dt}$, which is the tangent vector to a t-parameterized curve $C(t)$ given by the equations $u^i = u^i(t)$ where $i = 1, \cdots, n$, then the inner product:

$$A^i_{;r} \frac{du^r}{dt} \tag{397}$$

5.3 Absolute Differentiation

is a tensor of the same rank and type as the tensor A^i. The tensor given by the expression of Eq. 397 is called the "absolute" or "intrinsic" or "absolute covariant" derivative of the tensor A^i along the curve C and is symbolized by $\frac{\delta A^i}{\delta t}$, that is:

$$\frac{\delta A^i}{\delta t} = A^i_{;r} \frac{du^r}{dt} \tag{398}$$

In fact, instead of basing the definition of absolute differentiation on the definition of the covariant differentiation as a dot product of the covariant derivative with the tangent to the curve as seen above, we can define absolute differentiation independently by following the same method that we used to obtain the covariant derivative through applying the differentiation process on the basis vectors as well as the components of the tensor (as seen in the derivation of Eqs. 358-360) and hence the expressions of the absolute derivative (e.g. the upcoming Eqs. 401 and 402) can be obtained directly by applying total differentiation to the tensor including its basis vectors. For example, the absolute derivative of a covariant vector $\mathbf{A} = A^i \mathbf{E}_i$ can be obtained as follows:

$$\begin{aligned}
\frac{d\mathbf{A}}{dt} &= \frac{d}{dt}\left(A^i \mathbf{E}_i\right) & & (399) \\
&= \mathbf{E}_i \frac{dA^i}{dt} + A^i \frac{d\mathbf{E}_i}{dt} & & \text{(product rule)} \\
&= \mathbf{E}_i \frac{dA^i}{dt} + A^i \frac{\partial \mathbf{E}_i}{\partial u^j} \frac{du^j}{dt} & & \text{(chain rule)} \\
&= \mathbf{E}_i \frac{dA^i}{dt} + A^i \Gamma^k_{ij} \mathbf{E}_k \frac{du^j}{dt} & & \text{(Eq. 312)} \\
&= \mathbf{E}_i \frac{dA^i}{dt} + \mathbf{E}_i A^k \Gamma^i_{kj} \frac{du^j}{dt} & & \text{(exchanging dummy indices } i, k\text{)} \\
&= \left(\frac{dA^i}{dt} + A^k \Gamma^i_{kj} \frac{du^j}{dt}\right) \mathbf{E}_i \\
&= \frac{\delta A^i}{\delta t} \mathbf{E}_i & & \text{(definition of intrinsic derivative)}
\end{aligned}$$

which is the same as the upcoming Eq. 401.

Following the above definitions, the absolute derivative of a differentiable scalar f is the same as the ordinary total derivative, that is:

$$\frac{\delta f}{\delta t} = \frac{df}{dt} \tag{400}$$

This is because the covariant derivative of a scalar is the same as the ordinary partial derivative or because a scalar has no association with basis vectors to differentiate. The absolute derivative of a differentiable contravariant vector A^i with respect to the parameter t is given by:

$$\frac{\delta A^i}{\delta t} = \frac{dA^i}{dt} + \Gamma^i_{kj} A^k \frac{du^j}{dt} \tag{401}$$

5.3 Absolute Differentiation

Similarly, for a differentiable covariant vector A_i we have:

$$\frac{\delta A_i}{\delta t} = \frac{dA_i}{dt} - \Gamma^k_{ij} A_k \frac{du^j}{dt} \qquad (402)$$

Absolute differentiation can be easily extended to higher rank (> 1) differentiable tensors of type (m, n) along parameterized curves following the given examples and the pattern of covariant derivative. For instance, the absolute derivative of a mixed tensor of type $(1, 2)$ A^i_{jk} along a t-parameterized curve C is given by:

$$\frac{\delta A^i_{jk}}{\delta t} = A^i_{jk;b} \frac{du^b}{dt} = \frac{dA^i_{jk}}{dt} + \Gamma^i_{ab} A^a_{jk} \frac{du^b}{dt} - \Gamma^a_{jb} A^i_{ak} \frac{du^b}{dt} - \Gamma^a_{bk} A^i_{ja} \frac{du^b}{dt} \qquad (403)$$

Since the absolute derivative is given generically by:[26]

$$\frac{\delta \mathbf{A}}{\delta t} = \mathbf{A}_{;k} \frac{du^k}{dt} \qquad (404)$$

it can be seen as an instance of the chain rule of differentiation where the two contracted indices represent the in-between coordinate differential. This can also be concluded from the above method of derivation of Eq. 399. Because the absolute derivative along a curve is just an inner product of the covariant derivative with the tangent vector to the curve, the well known rules of ordinary differentiation of sums and products also apply to absolute differentiation, as for covariant differentiation, that is:

$$\frac{\delta}{\delta t}(a\mathbf{A} + b\mathbf{B}) = a\frac{\delta \mathbf{A}}{\delta t} + b\frac{\delta \mathbf{B}}{\delta t} \qquad (405)$$

$$\frac{\delta}{\delta t}(\mathbf{A} \circ \mathbf{B}) = \left(\frac{\delta \mathbf{A}}{\delta t} \circ \mathbf{B}\right) + \left(\mathbf{A} \circ \frac{\delta \mathbf{B}}{\delta t}\right) \qquad (406)$$

where a and b are constant scalars, $\mathbf{A}$ and $\mathbf{B}$ are differentiable tensors and the symbol $\circ$ denotes an inner or outer product of tensors. However, the order of the tensors in the products should be observed since tensor multiplication is not commutative.

Because absolute differentiation follows the style of covariant differentiation, the metric tensor in its different variance types is in lieu of a constant with respect to absolute differentiation, that is:

$$\frac{\delta g_{ij}}{\delta t} = 0 \qquad \frac{\delta g^{ij}}{\delta t} = 0 \qquad (407)$$

and hence it passes through the absolute derivative operator, that is:

$$\frac{\delta (g_{ij} A^j)}{\delta t} = g_{ij} \frac{\delta A^j}{\delta t} \qquad \frac{\delta (g^{ij} A_j)}{\delta t} = g^{ij} \frac{\delta A_j}{\delta t} \qquad (408)$$

---

[26] We use the absolute derivative notation with the symbolic notation of tensors in this equation and some of the upcoming equations to ease the notation. This is similar to the use of the covariant derivative notation with the symbolic notation as seen here and in previous equations. The meaning of these notations should be clear

5.3 Absolute Differentiation

For coordinate systems in which all the components of the metric tensor are constants, the absolute derivative is the same as the ordinary total derivative, as it is the case in rectilinear coordinate systems, because the Christoffel symbol terms are zero in these systems according to Eq. 308. The absolute derivative of a tensor along a given curve is unique, and hence the ordinary derivative of the tensor along that curve in a rectangular coordinate system is the same as the absolute derivative of the tensor along that curve in any other system although the forms in the two systems may be different.

To sum up, we list in the following bullet points the main rules of covariant and absolute differentiation:
1. Tensor differentiation (represented by covariant and absolute differentiation) is the same as ordinary differentiation (represented by partial and total differentiation) but with the application of the differentiation process not only on the tensor components but also on the basis vectors that associate these components.
2. The sum and product rules of differentiation apply to covariant and absolute differentiation as for ordinary differentiation (i.e. partial and total).
3. The covariant and absolute derivatives of tensors are tensors.
4. The covariant and absolute derivatives of scalars and affine tensors of higher ranks are the same as the ordinary derivatives.
5. The covariant and absolute derivative operators commute with the contraction of indices.
6. The covariant and absolute derivatives of the metric, Kronecker and permutation tensors as well as the basis vectors vanish identically in any coordinate system, that is:

$$g_{ij|q} = 0 \qquad g^{ij}_{\ |q} = 0 \qquad (409)$$

$$\delta^i_{j|q} = 0 \qquad \delta^{ij}_{kl|q} = 0 \qquad (410)$$

$$\underline{\epsilon}_{ijk|q} = 0 \qquad \underline{\epsilon}^{ijk}_{\ |q} = 0 \qquad (411)$$

$$\mathbf{E}_{i|q} = \mathbf{0} \qquad \mathbf{E}^i_{\ |q} = \mathbf{0} \qquad (412)$$

where the sign | represents covariant or absolute differentiation with respect to the space coordinate u^q. Hence, these tensors should be treated like constants in tensor differentiation. This applies to the covariant and contravariant forms of these tensors, as well as the mixed form when it is applicable. It also applies to the generalized, as well as the ordinary, Kronecker delta as seen in the above equations.
7. Covariant differentiation increases the covariant rank of the differentiated tensor by one, while absolute differentiation does not change the rank or type of the differentiated tensor.
8. The covariant and absolute derivatives in rectilinear systems are the same as the partial and total derivatives respectively for all tensor ranks.

5.4 Exercises and Revision

5.1 Why tensor differentiation (represented by covariant and absolute derivatives) is needed in general coordinate systems to replace the ordinary differentiation (represented by partial and total derivatives)?

5.2 Show that in general coordinate systems, the ordinary differentiation of the components of non-scalar tensors with respect to the coordinates will not produce a tensor in general.

5.3 "The Christoffel symbols are affine tensors but not tensors". Explain and justify this statement.

5.4 What is the difference between the first and second kinds of the Christoffel symbols?

5.5 Show that the Christoffel symbols of both kinds are not general tensors by giving examples of these symbols being vanishing in some systems but not in other systems and considering the universality of the zero tensor (see § 3.1.4).

5.6 State the mathematical definitions of the Christoffel symbols of the first and second kinds. How these two kinds are transformed from each other?

5.7 What is the significance of the Christoffel symbols being solely dependent on the coefficients of the metric tensor in their relation to the underlying space and coordinate system?

5.8 Do the Christoffel symbols represent a property of the space, a property of the coordinate system, or a property of both?

5.9 If some of the Christoffel symbols vanish in a particular curvilinear coordinate system, should these some necessarily vanish in other curvilinear coordinate systems? Justify your answer by giving some examples.

5.10 Verify that the Christoffel symbols of the first and second kind are symmetric in their paired indices by using their mathematical definitions.

5.11 Correct, if necessary, the following equations:

$$\partial_j \mathbf{E}_i = -\Gamma_{ij}^k \mathbf{E}_k \qquad \partial_j \mathbf{E}^i = -\Gamma_{mj}^i \mathbf{E}^m$$

5.12 What is the significance of the following equations?

$$\mathbf{E}^k \cdot \partial_j \mathbf{E}_i = \Gamma_{ij}^k \qquad \mathbf{E}_k \cdot \partial_j \mathbf{E}^i = -\Gamma_{kj}^i \qquad \mathbf{E}_k \cdot \partial_j \mathbf{E}_i = [ij, k]$$

5.13 Derive the following relations giving full explanation of each step:

$$\partial_j g_{il} = [ij, l] + [lj, i] \qquad \Gamma_{ji}^j = \partial_i \left(\ln \sqrt{g} \right)$$

5.14 Assuming an orthogonal coordinate system, verify the following relation: $[ij, k] = 0$ where $i \neq j \neq k$.

5.15 Assuming an orthogonal coordinate system, verify the following relation: $\Gamma_{ji}^i = \frac{1}{2} \partial_j \ln g_{ii}$ with no sum over i.

5.16 Considering the identicality and difference of the indices of the Christoffel symbols of either kind, how many cases we have? List these cases.

5.17 Prove the following relation which is used in Eq. 325: $\partial_i g = g g^{jl} \partial_i g_{jl}$.

5.4 Exercises and Revision

5.18 In orthogonal coordinate systems of a 3D space the number of independent non-identically vanishing Christoffel symbols of either kind is only 15. Explain why.

5.19 Verify the following equations related to the Christoffel symbols in orthogonal coordinate systems in a 3D space:

$$[12,1] = h_1 h_{1,2} \qquad \Gamma^3_{23} = \frac{h_{3,2}}{h_3}$$

5.20 Justify the following statement: "In any coordinate system, all the Christoffel symbols of either kind vanish identically *iff* all the components of the metric tensor in the given coordinate system are constants".

5.21 Using Eq. 307 with Eq. 239, find the Christoffel symbols of the first kind corresponding to the Euclidean metric of cylindrical coordinate systems.

5.22 Give all the Christoffel symbols of the first and second kind of the following coordinate systems: orthonormal Cartesian, cylindrical and spherical.

5.23 Mention two important properties of the Christoffel symbols of either kind with regard to the order and similarity of their indices.

5.24 Using the entries in Eq. 336 and Table 1 and the properties of the Christoffel symbols of the second kind, derive these symbols corresponding to the metrics of the coordinate systems of question 5.22.

5.25 Write the following Christoffel symbols in terms of the coordinates instead of the indices assuming a cylindrical system: $[12,1]$, $[23,1]$, Γ^2_{21} and Γ^3_{32}. Do the same assuming a spherical system.

5.26 Show that all the Christoffel symbols will vanish when the components of the metric tensor are constants.

5.27 Why the Christoffel symbols of either kind may be superscripted or subscripted by the symbol of the underlying metric tensor? When this (or other measures for indicating the underlying metric tensor) becomes necessary? Mention some of the other measures used to indicate the underlying metric tensor.

5.28 Explain why the total number of independent Christoffel symbols of each kind is equal to $\frac{n^2(n+1)}{2}$.

5.29 Why covariant differentiation of tensors is regarded as a generalization of the ordinary partial differentiation?

5.30 In general curvilinear coordinate systems, the variation of the basis vectors should also be considered in the differentiation process of non-scalar tensors. Why?

5.31 State the mathematical definition of contravariant differentiation of a tensor A_i.

5.32 Obtain analytical expressions for $A_{i;j}$ and $B^i_{;j}$ by differentiating the vectors $\mathbf{A} = A_i \mathbf{E}^i$ and $\mathbf{B} = B^i \mathbf{E}_i$.

5.33 Repeat question 5.32 with the rank-2 tensors $\mathbf{C} = C_{ij} \mathbf{E}^i \mathbf{E}^j$ and $\mathbf{D} = D^{ij} \mathbf{E}_i \mathbf{E}_j$ to obtain $C_{ij;k}$ and $D^{ij}_{;k}$.

5.34 For a differentiable tensor $\mathbf{A}$ of type (m,n), the covariant derivative with respect to the coordinate u^k is given by:

$$A^{i_1 i_2 \ldots i_m}_{j_1 j_2 \ldots j_n ; k} = \frac{\partial A^{i_1 i_2 \ldots i_m}_{j_1 j_2 \ldots j_n}}{\partial u^k} + \Gamma^{i_1}_{lk} A^{l i_2 \ldots i_m}_{j_1 j_2 \ldots j_n} + \Gamma^{i_2}_{lk} A^{i_1 l \ldots i_m}_{j_1 j_2 \ldots j_n} + \cdots + \Gamma^{i_m}_{lk} A^{i_1 i_2 \ldots l}_{j_1 j_2 \ldots j_n}$$

5.4 Exercises and Revision

$$-\Gamma^l_{j_1k}A^{i_1i_2...i_m}_{lj_2...j_n} - \Gamma^l_{j_2k}A^{i_1i_2...i_m}_{j_1l...j_n} - \cdots - \Gamma^l_{j_nk}A^{i_1i_2...i_m}_{j_1j_2...l}$$

Extract from the pattern of this expression the practical rules that should be followed in writing the analytical expressions of covariant derivative of tensors of any rank and type.

5.35 In the expression of covariant derivative, what the partial derivative term stands for and what the Christoffel symbol terms represent?

5.36 For the covariant derivative of a type (m, n, w) tensor, obtain the number of total terms, the number of negative Christoffel symbol terms and the number of positive Christoffel symbol terms.

5.37 What is the rank and type of the covariant derivative of a tensor of rank-n and type (p, q)?

5.38 The covariant derivative of a differentiable scalar function is the same as the ordinary partial derivative. Why?

5.39 What is the significance of the dependence of the covariant derivative on the Christoffel symbols with regard to its relation to the space and coordinate system?

5.40 The covariant derivative of tensors in coordinate systems with constant basis vectors is the same as the ordinary partial derivative for all tensor ranks. Why?

5.41 Express, mathematically, the fact that the metric tensor is in lieu of constant with respect to covariant differentiation.

5.42 Which rules of ordinary partial differentiation also apply to covariant differentiation and which rules do not? State all these rules symbolically for both ordinary and covariant differentiation.

5.43 Explain why the covariant differential operators with respect to different indices do not commute, i.e. $\partial_{;i}\partial_{;j} \neq \partial_{;j}\partial_{;i}$ $(i \neq j)$.

5.44 State the Ricci theorem about covariant differentiation of the metric tensor and prove it with full justification of each step.

5.45 State, symbolically, the commutative property of the covariant derivative operator with the index shifting operator (which is based on the Ricci theorem) using the symbolic notation one time and the indicial notation another.

5.46 Verify that the ordinary Kronecker delta tensor is constant with respect to covariant differentiation.

5.47 State, symbolically, the fact that covariant differentiation and contraction of index operations commute with each other.

5.48 What is the condition on the components of the metric tensor that makes the covariant derivative become ordinary partial derivative for all tensor ranks?

5.49 Prove that covariant differentiation and contraction of indices commute.

5.50 What is the mathematical condition that is required if the mixed second order partial derivatives should be equal, i.e. $\partial_i\partial_j = \partial_j\partial_i$ $(i \neq j)$?

5.51 What is the mathematical condition that is required if the mixed second order covariant derivatives should be equal, i.e. $\partial_{;i}\partial_{;j} = \partial_{;j}\partial_{;i}$ $(i \neq j)$?

5.52 Derive analytical expressions for $A_{i;jk}$ and $A_{i;kj}$ and hence verify that $A_{i;jk} \neq A_{i;kj}$.

5.53 From the result of exercise 5.52 plus Eq. 560, verify the following relation: $A_{i;jk} -$

5.4 Exercises and Revision

$$A_{i;kj} = A_a R^a{}_{ijk}.$$

5.54 What is the covariant derivative of a relative scalar f of weight w? What is the covariant derivative of a rank-2 relative tensor A^i_j of weight w?

5.55 Why the covariant derivative of a non-scalar tensor with constant components is not necessarily zero in general coordinate systems? Which term of the covariant derivative of such a tensor will vanish?

5.56 Show that: $A_{i;j} = A_{j;i}$ where **A** is a gradient of a scalar field.

5.57 Show that the covariant derivative of the basis vectors of the covariant and contravariant types is identically zero, i.e. $\mathbf{E}_{i;j} = \mathbf{0}$ and $\mathbf{E}^i_{;j} = \mathbf{0}$.

5.58 Prove the following identity:

$$\partial_k \left(g_{ij} A^i B^j \right) = A_{i;k} B^i + A^i B_{i;k}$$

5.59 Define absolute differentiation descriptively and mathematically. What are the other names of absolute derivative?

5.60 Write the mathematical expression for the absolute derivative of the tensor field $A^{ij}{}_k$ which is defined over a space curve $C(t)$.

5.61 Why the absolute derivative of a differentiable scalar is the same as its ordinary total derivative, i.e. $\frac{\delta f}{\delta t} = \frac{df}{dt}$?

5.62 Why the absolute derivative of a differentiable non-scalar tensor is the same as its ordinary total derivative in rectilinear coordinate systems?

5.63 From the pattern of covariant derivative of a general tensor, obtain the pattern of its absolute derivative.

5.64 We have $\mathbf{A} = A^{ij}{}_k \mathbf{E}_i \mathbf{E}_j \mathbf{E}^k$. Apply the ordinary total differentiation process (i.e. $\frac{d\mathbf{A}}{dt}$) onto this tensor (including its basis vectors) to obtain its absolute derivative.

5.65 Which rules of ordinary total differentiation also apply to intrinsic differentiation and which rules do not? State all these rules symbolically for both ordinary and intrinsic differentiation.

5.66 Using your knowledge about covariant differentiation and the fact that absolute differentiation follows the style of covariant differentiation, obtain all the rules of absolute differentiation of the metric tensor, the Kronecker delta tensor and the index shifting and index replacement operators. Express all these rules in words and in symbols.

5.67 Justify the following statement: "For coordinate systems in which all the components of the metric tensor are constants, the absolute derivative is the same as the ordinary total derivative".

5.68 The absolute derivative of a tensor along a given curve is unique. What this means?

5.69 Summarize all the main properties and rules that govern covariant and absolute differentiation.

Chapter 6
Differential Operations

In this chapter, we examine the main differential operations which are based on the nabla operator ∇ as defined in tensor calculus using largely tensor notation. These operations are based on the various types of interaction between the vector differential operator nabla and tensors of different ranks where some of these interactions involve the dot and cross product operations. The chapter will investigate these operations in general coordinate systems and in general orthogonal coordinate systems which are a special case of the general coordinate systems. The chapter will also investigate these operations in Cartesian coordinate systems as well as the two most important and widely used curvilinear orthogonal coordinate systems, namely the cylindrical and spherical systems, because of their particular importance and widespread application in science and engineering. Due to the likely familiarity of the reader with these operations in Cartesian systems, which is usually acquired at this level from a previous course on vector calculus, we start our investigation from the Cartesian system. This may help to remind the reader and make the subsequent sections easier to understand.

Regarding the cylindrical and spherical coordinate systems, we can use indexed general coordinates like u^1, u^2 and u^3 to represent the cylindrical coordinates (ρ, ϕ, z) and the spherical coordinates (r, θ, ϕ) and hence we can express these operations in tensor notation as we do for the other systems. However, for the sake of clarity and to follow the more conventional practice, we use the coordinates of these systems as suffixes in place of the usual indices used in the tensor notation. In this context, we should insist that these suffixes are labels and not indices and therefore they do not follow the rules and conventions of tensor indices such as following the summation convention. In fact, there is another reason for the use of suffix labels instead of symbolic indices that is the components in the cylindrical and spherical coordinates are physical, not covariant or contravariant, and hence suffixing with coordinates is more appropriate (see § 3.3).

Before we start this investigation, we should remark that the differentiation of a tensor increases its rank by one, by introducing an extra covariant index, unless it implies a contraction in which case it reduces the rank by one. Therefore the gradient of a scalar is a vector and the gradient of a vector is a rank-2 tensor, while the divergence of a vector is a scalar and the divergence of a rank-2 tensor is a vector. This may be justified by the fact that the gradient operator is a vector operator. On the other hand the Laplacian operator does not change the rank since it is a scalar operator; hence the Laplacian of a scalar is a scalar and the Laplacian of a vector is a vector and so on.

We should also remark that there are other nabla based operators and operations which are subsidiary to the main ones and are used in pure and applied mathematics and science. For example, the following scalar differential operator, defined in Cartesian coordinates, is commonly used in science such as fluid dynamics:

$$\mathbf{A} \cdot \nabla = A_i \nabla_i = A_i \frac{\partial}{\partial x_i} = A_i \partial_i \tag{413}$$

where $\mathbf{A}$ is a vector. As explained earlier (see § 1.2), the order of A_i and ∂_i should be respected. Another example is the following vector differential operator which also has common applications in science:

$$[\mathbf{A} \times \nabla]_i = \epsilon_{ijk} A_j \partial_k \tag{414}$$

where, again, the order should be respected and the operator is defined in Cartesian coordinates.

6.1 Cartesian Coordinate System

6.1.1 Operators

The nabla vector operator ∇ is a spatial partial differential operator which is defined in Cartesian coordinate systems by:

$$\nabla_i = \frac{\partial}{\partial x_i} \tag{415}$$

Similarly, the Laplacian scalar operator is given by:

$$\nabla^2 = \delta_{ij} \frac{\partial^2}{\partial x_i \partial x_j} = \frac{\partial^2}{\partial x_i \partial x_i} = \nabla_{ii} = \partial_{ii} \tag{416}$$

We note that the Laplacian operator may also be notated with Δ (as well as several other symbols such as ∂_{ii}^2). However, in the present book we do not use this notation for this purpose.

6.1.2 Gradient

Based on the above definition of nabla, the gradient of a differentiable scalar function of position f is a vector obtained by applying the nabla operator to f and hence it is defined by:

$$[\nabla f]_i = \nabla_i f = \frac{\partial f}{\partial x_i} = \partial_i f = f_{,i} \tag{417}$$

Similarly, the gradient of a differentiable vector field $\mathbf{A}$ is the outer product (refer to § 3.2.3) between the nabla operator and the vector and hence it is a rank-2 tensor, that is:

$$[\nabla \mathbf{A}]_{ij} = \partial_i A_j \tag{418}$$

This definition can be easily extended to higher rank tensors.

6.1.3 Divergence

The divergence of a differentiable vector $\mathbf{A}$ is the dot product of the nabla operator and the vector $\mathbf{A}$ and hence it is a scalar given by:

$$\nabla \cdot \mathbf{A} = \delta_{ij} \frac{\partial A_i}{\partial x_j} = \frac{\partial A_i}{\partial x_i} = \nabla_i A_i = \partial_i A_i = A_{i,i} \tag{419}$$

The divergence operation can also be viewed as taking the gradient of the vector followed by a contraction. Hence, the divergence of a vector is invariant because it is the trace of a rank-2 tensor (see § 7.1.2). It may also be argued more simply that the divergence of a vector is a scalar and hence it is invariant.

Similarly, the divergence of a differentiable rank-2 tensor $\mathbf{A}$ is a vector defined in one of its forms by:

$$[\nabla \cdot \mathbf{A}]_i = \partial_j A_{ji} \tag{420}$$

and in another form by:

$$[\nabla \cdot \mathbf{A}]_j = \partial_i A_{ji} \tag{421}$$

These two forms may be given respectively, using the symbolic notation, by:

$$\nabla \cdot \mathbf{A} \qquad \text{and} \qquad \nabla \cdot \mathbf{A}^T \tag{422}$$

where $\mathbf{A}^T$ is the transpose of $\mathbf{A}$. More generally, the divergence of a tensor of rank $n \geq 2$, which is a tensor of rank-$(n-1)$, can be defined in several forms, which are different in general, depending on the choice of the contracted index.

6.1.4 Curl

The curl of a differentiable vector $\mathbf{A}$ is the cross product of the nabla operator and the vector $\mathbf{A}$ and hence it is a vector defined by:

$$[\nabla \times \mathbf{A}]_i = \epsilon_{ijk} \frac{\partial A_k}{\partial x_j} = \epsilon_{ijk} \nabla_j A_k = \epsilon_{ijk} \partial_j A_k = \epsilon_{ijk} A_{k,j} \tag{423}$$

The curl operation may be generalized to tensors of rank > 1 (see for example § 7.1.4), and hence the curl of a differentiable rank-2 tensor $\mathbf{A}$ can be defined as a rank-2 tensor given by:

$$[\nabla \times \mathbf{A}]_{ij} = \epsilon_{imn} \partial_m A_{nj} \tag{424}$$

The last example can be easily extended to higher rank tensors. We note that there is more than one possibility for the contraction of the last index of the permutation tensor with one of the tensor indices when the rank is > 1.

6.1.5 Laplacian

The Laplacian of a differentiable scalar f is given by:

$$\nabla^2 f = \delta_{ij} \frac{\partial^2 f}{\partial x_i \partial x_j} = \frac{\partial^2 f}{\partial x_i \partial x_i} = \nabla_{ii} f = \partial_{ii} f = f_{,ii} \qquad (425)$$

The Laplacian of a differentiable vector $\mathbf{A}$ is defined for each component of the vector in a similar manner to the definition of the Laplacian acting on a scalar, that is:

$$\left[\nabla^2 \mathbf{A}\right]_i = \nabla^2 \left[\mathbf{A}\right]_i = \partial_{jj} A_i \qquad (426)$$

This definition can be easily extended to higher rank tensors.

6.2 General Coordinate System

Here, we investigate the differential operators and operations in general coordinate systems. We note that the definitions of the differential operations in Cartesian systems, as given in the vector calculus texts and as outlined in § 6.1, are essentially valid in general non-Cartesian coordinate systems if the operations are extended to include the basis vectors as well as the components, as we will see in the following subsections. We also note that the analytical expressions of the differential operations can be obtained directly if the expression for the nabla operator ∇ and the spatial derivatives of the basis vectors of the given coordinate system are known.

6.2.1 Operators

The nabla operator ∇, which is a spatial partial differential vector operator, is defined in general coordinate systems by:

$$\nabla = \mathbf{E}^i \partial_i \qquad (427)$$

Similarly, the Laplacian operator is defined generically by:

$$\nabla^2 = \text{div grad} = \nabla \cdot \nabla \qquad (428)$$

More details about these operators will follow in the next subsections.

6.2.2 Gradient

Based on the definition of the nabla operator, as given in the previous subsection, the gradient of a differentiable scalar function of position, f, is given by:

$$\nabla f = \mathbf{E}^i \partial_i f = \mathbf{E}^i f_{,i} \qquad (429)$$

The components of this expression represent the covariant form of a rank-1 tensor, i.e. $[\nabla f]_i = f_{,i}$, as it should be since the gradient operation increases the covariant rank of a tensor by one, as indicated earlier. Since this expression consists of a contravariant basis

6.2.2 Gradient

vector and a covariant component, the gradient in general coordinate systems is invariant under admissible coordinate transformations.

The gradient of a differentiable covariant vector **A** can similarly be defined as follows:[27]

$$\nabla \mathbf{A} = \mathbf{E}^i \partial_i \left(A_j \mathbf{E}^j \right) \tag{430}$$
$$= \mathbf{E}^i \mathbf{E}^j \partial_i A_j + \mathbf{E}^i A_j \partial_i \mathbf{E}^j \qquad \text{(product rule)}$$
$$= \mathbf{E}^i \mathbf{E}^j \partial_i A_j + \mathbf{E}^i A_j \left(-\Gamma^j_{ki} \mathbf{E}^k \right) \qquad \text{(Eq. 313)}$$
$$= \mathbf{E}^i \mathbf{E}^j \partial_i A_j - \mathbf{E}^i \mathbf{E}^j \Gamma^k_{ji} A_k \qquad \text{(relabeling dummy indices } j \text{ \& } k\text{)}$$
$$= \mathbf{E}^i \mathbf{E}^j \left(\partial_i A_j - \Gamma^k_{ji} A_k \right) \qquad \text{(taking common factor } \mathbf{E}^i \mathbf{E}^j\text{)}$$
$$= \mathbf{E}^i \mathbf{E}^j A_{j;i} \qquad \text{(definition of covariant derivative)}$$

and hence it is the covariant derivative of the vector. Similarly, for a differentiable contravariant vector **A** the gradient is given by:

$$\nabla \mathbf{A} = \mathbf{E}^i \partial_i \left(A^j \mathbf{E}_j \right) \tag{431}$$
$$= \mathbf{E}^i \mathbf{E}_j \partial_i A^j + \mathbf{E}^i A^j \partial_i \mathbf{E}_j \qquad \text{(product rule)}$$
$$= \mathbf{E}^i \mathbf{E}_j \partial_i A^j + \mathbf{E}^i A^j \left(\Gamma^k_{ji} \mathbf{E}_k \right) \qquad \text{(Eq. 312)}$$
$$= \mathbf{E}^i \mathbf{E}_j \partial_i A^j + \mathbf{E}^i \mathbf{E}_j \Gamma^j_{ki} A^k \qquad \text{(relabeling dummy indices } j \text{ \& } k\text{)}$$
$$= \mathbf{E}^i \mathbf{E}_j \left(\partial_i A^j + \Gamma^j_{ki} A^k \right) \qquad \text{(taking common factor } \mathbf{E}^i \mathbf{E}_j\text{)}$$
$$= \mathbf{E}^i \mathbf{E}_j A^j_{;i} \qquad \text{(definition of covariant derivative)}$$

The components of the gradients of covariant and contravariant vectors represent, respectively, the covariant and mixed forms of a rank-2 tensor, as they should be since the gradient operation increases the covariant rank of a tensor by one.

The gradient of higher rank tensors is similarly defined. For example, the gradient of a rank-2 tensor in its covariant, contravariant and mixed form is given by:

$$\nabla \mathbf{A} = \mathbf{E}^i \mathbf{E}^j \mathbf{E}^k \left(\partial_i A_{jk} - \Gamma^l_{ji} A_{lk} - \Gamma^l_{ki} A_{jl} \right) = \mathbf{E}^i \mathbf{E}^j \mathbf{E}^k A_{jk;i} \qquad \text{(covariant)} \tag{432}$$
$$\nabla \mathbf{A} = \mathbf{E}^i \mathbf{E}_j \mathbf{E}_k \left(\partial_i A^{jk} + \Gamma^j_{li} A^{lk} + \Gamma^k_{li} A^{jl} \right) = \mathbf{E}^i \mathbf{E}_j \mathbf{E}_k A^{jk}_{;i} \qquad \text{(contravariant)} \tag{433}$$
$$\nabla \mathbf{A} = \mathbf{E}^i \mathbf{E}^j \mathbf{E}_k \left(\partial_i A_j^{\ k} - \Gamma^l_{ji} A_l^{\ k} + \Gamma^k_{li} A_j^{\ l} \right) = \mathbf{E}^i \mathbf{E}^j \mathbf{E}_k A_j^{\ k}_{;i} \qquad \text{(mixed)} \tag{434}$$
$$\nabla \mathbf{A} = \mathbf{E}^i \mathbf{E}_j \mathbf{E}^k \left(\partial_i A^j_{\ k} + \Gamma^j_{li} A^l_{\ k} - \Gamma^l_{ki} A^j_{\ l} \right) = \mathbf{E}^i \mathbf{E}_j \mathbf{E}^k A^j_{\ k;i} \qquad \text{(mixed)} \tag{435}$$

We finally remark that the contravariant form of the gradient operation can be obtained by using the index raising operator. For example, the contravariant form of the gradient of a scalar f is given by:

$$[\nabla f]^i = \partial^i f = g^{ij} \partial_j f = g^{ij} f_{,j} = f^{,i} \tag{436}$$

---

[27] We note that the basis vector $\mathbf{E}^i$ which associates the derivative operator in the following equations (as well as in similar equations and expressions) should be the last one in the basis tensor so that the order of the indices in the components and in the basis tensor are the same. So, strictly we should write $\nabla \mathbf{A} = \partial_i \left(A_j \mathbf{E}^j \right) \mathbf{E}^i$ where ∂_i acts only on what is inside the parentheses and hence the final expression becomes $\nabla \mathbf{A} = \mathbf{E}^j \mathbf{E}^i A_{j;i}$. However, to avoid confusion we put the vector to the left relying on this understanding.

6.2.3 Divergence

This can be easily extended to higher rank tensors. As seen, the contravariant form of the gradient operation increases the contravariant rank of the tensor by one.

6.2.3 Divergence

Generically, the divergence of a differentiable contravariant vector $\mathbf{A}$ is defined as follows:

$$\nabla \cdot \mathbf{A} = \mathbf{E}^i \partial_i \cdot \left(A^j \mathbf{E}_j \right) \tag{437}$$
$$= \mathbf{E}^i \cdot \partial_i \left(A^j \mathbf{E}_j \right)$$
$$= \mathbf{E}^i \cdot \left(A^j_{;i} \mathbf{E}_j \right) \qquad \text{(definition of covariant derivative)}$$
$$= \left(\mathbf{E}^i \cdot \mathbf{E}_j \right) A^j_{;i}$$
$$= \delta^i_j A^j_{;i} \qquad \text{(Eq. 215)}$$
$$= A^i_{;i} \qquad \text{(Eq. 172)}$$

In more details, the divergence of a differentiable contravariant vector A^i is a scalar obtained by contracting the covariant derivative index with the contravariant index of the vector, and hence:

$$\nabla \cdot \mathbf{A} = A^i_{;i} \tag{438}$$
$$= \partial_i A^i + \Gamma^i_{ji} A^j \qquad \text{(Eq. 362)}$$
$$= \partial_i A^i + A^j \frac{1}{\sqrt{g}} \partial_j \left(\sqrt{g} \right) \qquad \text{(Eq. 324)}$$
$$= \partial_i A^i + A^i \frac{1}{\sqrt{g}} \partial_i \left(\sqrt{g} \right) \qquad \text{(renaming dummy index } j\text{)}$$
$$= \frac{1}{\sqrt{g}} \partial_i \left(\sqrt{g} A^i \right) \qquad \text{(product rule)}$$

where g is the determinant of the covariant metric tensor g_{ij}. The last equality may be called the Voss-Weyl formula.

The divergence can also be obtained by raising the first index of the covariant derivative of a covariant vector using a contracting contravariant metric tensor, that is:

$$g^{ji} A_{j;i} = \left(g^{ji} A_j \right)_{;i} \qquad \text{(Eq. 376)} \tag{439}$$
$$= \left(A^i \right)_{;i} \qquad \text{(Eq. 228)}$$
$$= A^i_{;i}$$
$$= \nabla \cdot \mathbf{A} \qquad \text{(Eq. 438)}$$

as before. Accordingly, the divergence of a covariant vector A_j is obtained by using the raising operator, that is:

$$A^i_{;i} = g^{ij} A_{j;i} \tag{440}$$

6.2.4 Curl

For a rank-2 contravariant tensor **A**, the divergence is generically defined by:

$$\nabla \cdot \mathbf{A} = \mathbf{E}^i \partial_i \cdot \left(A^{jk} \mathbf{E}_j \mathbf{E}_k \right) \tag{441}$$
$$= \mathbf{E}^i \cdot \partial_i \left(A^{jk} \mathbf{E}_j \mathbf{E}_k \right)$$
$$= \mathbf{E}^i \cdot \left(A^{jk}_{;i} \mathbf{E}_j \mathbf{E}_k \right) \qquad \text{(definition of covariant derivative)}$$
$$= \left(\mathbf{E}^i \cdot \mathbf{E}_j \right) \mathbf{E}_k A^{jk}_{;i}$$
$$= \delta^i_j \mathbf{E}_k A^{jk}_{;i} \qquad \text{(Eq. 215)}$$
$$= \mathbf{E}_k A^{ik}_{;i} \qquad \text{(Eq. 172)}$$

The components of this expression represent a contravariant vector, as it should be since the divergence operation reduces the contravariant rank of a tensor by one.

More generally, considering the tensor components, the divergence of a differentiable rank-2 contravariant tensor A^{ij} is a contravariant vector obtained by contracting the covariant derivative index with one of the contravariant indices, that is:

$$[\nabla \cdot \mathbf{A}]^j = A^{ij}_{;i} \qquad \text{or} \qquad [\nabla \cdot \mathbf{A}]^i = A^{ij}_{;j} \tag{442}$$

And for a rank-2 mixed tensor A^i_j we have:

$$[\nabla \cdot \mathbf{A}]_j = A^i_{j;i} \tag{443}$$

Similarly, for a general tensor of type (m,n): $\mathbf{A} = A^{i_1 i_2 \cdots i_k \cdots i_m}_{j_1 j_2 \cdots \cdots \cdots j_n}$, the divergence with respect to its kth contravariant index is defined by:

$$[\nabla \cdot \mathbf{A}]^{i_1 i_2 \cdots \cdots i_m}_{j_1 j_2 \cdots \cdots \cdots j_n} = \left(A^{i_1 i_2 \cdots s \cdots i_m}_{j_1 j_2 \cdots \cdots \cdots j_n} \right)_{;s} \tag{444}$$

with the absence of the contracted contravariant index i_k on the left hand side. As a matter of notation, it should be understood that $\nabla \cdot \mathbf{A}$ is lower than the original tensor **A** by just one contravariant index and hence, unlike the common use of this notation, it is not necessarily scalar.

6.2.4 Curl

The curl of a differentiable vector is the cross product of the nabla operator ∇ with the vector. For example, the curl of a vector **A** represented by covariant components is given by:

$$\operatorname{curl} \mathbf{A} = \nabla \times \mathbf{A} \tag{445}$$
$$= \mathbf{E}^i \partial_i \times A_j \mathbf{E}^j$$
$$= \mathbf{E}^i \times \partial_i \left(A_j \mathbf{E}^j \right)$$
$$= \mathbf{E}^i \times \left(A_{j;i} \mathbf{E}^j \right) \qquad \text{(Eq. 359)}$$
$$= A_{j;i} \left(\mathbf{E}^i \times \mathbf{E}^j \right)$$

6.2.5 Laplacian

$$= A_{j;i}\,\epsilon^{ijk}\mathbf{E}_k \qquad (\text{Eq. 274})$$

$$= \epsilon^{ijk} A_{j;i}\mathbf{E}_k$$

$$= \frac{\epsilon^{ijk}}{\sqrt{g}}\left(\partial_i A_j - \Gamma^l_{ji} A_l\right)\mathbf{E}_k \qquad (\text{Eqs. 166 \& 361})$$

Hence, the kth contravariant component of curl $\mathbf{A}$ is given by:

$$[\nabla \times \mathbf{A}]^k = \frac{\epsilon^{ijk}}{\sqrt{g}}\left(\partial_i A_j - \Gamma^l_{ji} A_l\right) \qquad (446)$$

On expanding the last equation for the three components of a vector in a 3D space, considering that the terms of the Christoffel symbols cancel out due to their symmetry in the two lower indices,[28] we obtain:

$$[\nabla \times \mathbf{A}]^1 = \frac{1}{\sqrt{g}}\left(\partial_2 A_3 - \partial_3 A_2\right) \qquad (447)$$

$$[\nabla \times \mathbf{A}]^2 = \frac{1}{\sqrt{g}}\left(\partial_3 A_1 - \partial_1 A_3\right) \qquad (448)$$

$$[\nabla \times \mathbf{A}]^3 = \frac{1}{\sqrt{g}}\left(\partial_1 A_2 - \partial_2 A_1\right) \qquad (449)$$

Hence, Eq. 446 will reduce to:

$$[\nabla \times \mathbf{A}]^k = \frac{\epsilon^{ijk}}{\sqrt{g}}\partial_i A_j \qquad (450)$$

6.2.5 Laplacian

Generically, the Laplacian of a differentiable scalar function of position, f, is defined as follows:

$$\nabla^2 f = \text{div}\left(\text{grad}\,f\right) = \nabla \cdot (\nabla f) \qquad (451)$$

Hence the simplest approach for obtaining the Laplacian in general coordinate systems is to insert the expression for the gradient, ∇f, into the expression for the divergence. However, because in general coordinate systems the divergence is defined only for tensors having at least one contravariant index to facilitate the contraction with the covariant

---

[28] That is:

$$\begin{aligned}
A_{i;j} - A_{j;i} &= \partial_j A_i - A_k \Gamma^k_{ij} - \partial_i A_j + A_k \Gamma^k_{ji} \\
&= \partial_j A_i - A_k \Gamma^k_{ij} - \partial_i A_j + A_k \Gamma^k_{ij} \\
&= \partial_j A_i - \partial_i A_j \\
&= A_{i,j} - A_{j,i}
\end{aligned}$$

6.2.5 Laplacian

derivative index (see for example Eqs. 437 and 444) whereas the gradient of a scalar is a covariant tensor, the index of the gradient should be raised first before applying the divergence operation, that is:

$$[\nabla f]^i = \partial^i f = g^{ij} \partial_j f \tag{452}$$

Now, according to Eq. 438 the divergence is given by:

$$\nabla \cdot \mathbf{A} = A^i_{;i} = \frac{1}{\sqrt{g}} \partial_i \left(\sqrt{g} A^i \right) \tag{453}$$

On defining $\mathbf{A} \equiv \mathbf{E}_i \partial^i f$ and replacing A^i in Eq. 453 with $\partial^i f$ using Eq. 452 we obtain:

$$\nabla^2 f = \nabla \cdot \left(\mathbf{E}_i \partial^i f \right) = \frac{1}{\sqrt{g}} \partial_i \left(\sqrt{g} g^{ij} \partial_j f \right) \tag{454}$$

which is the expression for the Laplacian of a scalar function f in general coordinate systems.

Another approach for developing the Laplacian expression in general coordinate systems is to apply the first principles by using the definitions and basic properties of the operations involved, that is:

$$\begin{aligned}
\nabla^2 f &= \nabla \cdot (\nabla f) & (455)\\
&= \mathbf{E}^i \partial_i \cdot \left(\mathbf{E}^j \partial_j f \right) \\
&= \mathbf{E}^i \cdot \partial_i \left(\mathbf{E}^j \partial_j f \right) \\
&= \mathbf{E}^i \cdot \partial_i \left(\mathbf{E}^j f_{,j} \right) \\
&= \mathbf{E}^i \cdot \left(\mathbf{E}^j f_{,j;i} \right) & \text{(Eq. 359)} \\
&= \left(\mathbf{E}^i \cdot \mathbf{E}^j \right) f_{,j;i} \\
&= g^{ij} f_{,j;i} & \text{(Eq. 214)} \\
&= \left(g^{ij} f_{,j} \right)_{;i} & \text{(Eq. 376)} \\
&= \left(g^{ij} \partial_j f \right)_{;i} \\
&= \partial_i \left(g^{ij} \partial_j f \right) + \left(g^{kj} \partial_j f \right) \Gamma^i_{ki} & \text{(Eq. 362)} \\
&= \partial_i \left(g^{ij} \partial_j f \right) + \left(g^{ij} \partial_j f \right) \Gamma^k_{ik} & \text{(renaming dummy indices } i \text{ \& } k\text{)} \\
&= \partial_i \left(g^{ij} \partial_j f \right) + \left(g^{ij} \partial_j f \right) \frac{1}{\sqrt{g}} \left(\partial_i \sqrt{g} \right) & \text{(Eq. 324)} \\
&= \frac{1}{\sqrt{g}} \left[\sqrt{g} \partial_i \left(g^{ij} \partial_j f \right) + \left(g^{ij} \partial_j f \right) \partial_i \sqrt{g} \right] & \text{(taking } \tfrac{1}{\sqrt{g}} \text{ factor)} \\
&= \frac{1}{\sqrt{g}} \partial_i \left(\sqrt{g} g^{ij} \partial_j f \right) & \text{(product rule)}
\end{aligned}$$

which is the same as before (see Eq. 454).

The Laplacian of a scalar f may also be shorthand notated with:

$$\nabla^2 f = g^{ij} f_{,ij} \tag{456}$$

The Laplacian of non-scalar tensors can be similarly defined. For example, the Laplacian of a vector $\mathbf{B}$ in its contravariant and covariant forms, B^i and B_i, is a vector $\mathbf{A}$ (i.e. $\mathbf{A} = \nabla^2 \mathbf{B}$) which may be defined in general coordinate systems as:

$$A^i = g^{jk} B^i{}_{;jk} \qquad\qquad A_i = g^{jk} B_{i;jk} \tag{457}$$

As indicated earlier, the Laplacian of a tensor is a tensor of the same rank and variance type.

6.3 Orthogonal Coordinate System

In this section, we state the main differential operators and operations in general orthogonal coordinate systems. These operators and operations are special cases of those of general coordinate systems which were derived in § 6.2. However, due to the widespread use of orthogonal coordinate systems, it is beneficial to state the most important of these operators and operations although they can be easily obtained from the formulae of general coordinate systems.

For clarity, general orthogonal coordinate systems are identified in this section by the coordinates $(q^1, \ldots, q^n)$ with unit basis vectors $(\mathbf{q}_1, \ldots, \mathbf{q}_n)$ and scale factors $(h_1, \ldots, h_n)$ where:

$$\mathbf{q}_i = \sum_j \frac{1}{h_i} \frac{\partial x^j}{\partial q^i} \mathbf{e}_j = \sum_j h_i \frac{\partial q^i}{\partial x^j} \mathbf{e}_j \qquad \text{(no sum on } i\text{)} \tag{458}$$

$$h_i = |\mathbf{E}_i| = \left|\frac{\partial \mathbf{r}}{\partial q^i}\right| = \left[\sum_j \left(\frac{\partial x^j}{\partial q^i}\right)^2\right]^{1/2} = \left[\sum_j \left(\frac{\partial q^i}{\partial x^j}\right)^2\right]^{-1/2} \tag{459}$$

In the last equations, x^j and $\mathbf{e}_j$ are respectively the coordinates and unit basis vectors in a Cartesian rectangular system, and $\mathbf{r} = x^i \mathbf{e}_i$ is the position vector in that system. We remark that in orthogonal coordinate systems the covariant and contravariant normalized basis vectors are identical, as established previously in § 3.3, and hence $\mathbf{q}^i = \mathbf{q}_i$ and $\mathbf{e}^j = \mathbf{e}_j$.

6.3.1 Operators

The nabla operator in general orthogonal coordinate systems is given by:

$$\nabla = \sum_i \frac{\mathbf{q}_i}{h_i} \frac{\partial}{\partial q^i} \tag{460}$$

while the Laplacian operator, assuming a 3D space, is given by:

$$\nabla^2 = \frac{1}{h_1 h_2 h_3} \sum_{i=1}^{3} \frac{\partial}{\partial q^i} \left(\frac{h_1 h_2 h_3}{(h_i)^2} \frac{\partial}{\partial q^i} \right) \tag{461}$$

6.3.2 Gradient

The gradient of a differentiable scalar f in orthogonal coordinate systems, assuming a 3D space, is given by:

$$\nabla f = \sum_{i=1}^{3} \frac{\mathbf{q}_i}{h_i} \frac{\partial f}{\partial q^i} = \frac{\mathbf{q}_1}{h_1} \frac{\partial f}{\partial q^1} + \frac{\mathbf{q}_2}{h_2} \frac{\partial f}{\partial q^2} + \frac{\mathbf{q}_3}{h_3} \frac{\partial f}{\partial q^3} \qquad (462)$$

6.3.3 Divergence

The divergence in orthogonal coordinate systems can be obtained from Eq. 438. Since for orthogonal coordinate systems the metric tensor according to Eqs. 233-236 is diagonal with $\sqrt{g} = h_1 h_2 h_3$ in a 3D space and $h_i A^i = \hat{A}^i$ (no summation) according to Eq. 144, the last line of Eq. 438 becomes:

$$\begin{aligned}
\nabla \cdot \mathbf{A} &= \frac{1}{\sqrt{g}} \frac{\partial}{\partial q^i} \left(\sqrt{g} A^i \right) \qquad (463) \\
&= \frac{1}{h_1 h_2 h_3} \sum_{i=1}^{3} \frac{\partial}{\partial q^i} \left(\frac{h_1 h_2 h_3}{h_i} \hat{A}^i \right) \\
&= \frac{1}{h_1 h_2 h_3} \left[\frac{\partial}{\partial q^1} \left(h_2 h_3 \hat{A}_1 \right) + \frac{\partial}{\partial q^2} \left(h_1 h_3 \hat{A}_2 \right) + \frac{\partial}{\partial q^3} \left(h_1 h_2 \hat{A}_3 \right) \right]
\end{aligned}$$

where $\mathbf{A}$ is a contravariant differentiable vector and $\hat{A}^i$ represents its physical components. This equation is the divergence of a vector in general orthogonal coordinates as defined in vector calculus. We note that in orthogonal coordinate systems the physical components are the same for covariant and contravariant forms, as established before in § 3.3, and hence $\hat{A}_i = \hat{A}^i$.

6.3.4 Curl

The curl of a differentiable vector $\mathbf{A}$ in orthogonal coordinate systems in 3D spaces is given by:

$$\nabla \times \mathbf{A} = \frac{1}{h_1 h_2 h_3} \begin{vmatrix} h_1 \mathbf{q}_1 & h_2 \mathbf{q}_2 & h_3 \mathbf{q}_3 \\ \frac{\partial}{\partial q^1} & \frac{\partial}{\partial q^2} & \frac{\partial}{\partial q^3} \\ h_1 \hat{A}_1 & h_2 \hat{A}_2 & h_3 \hat{A}_3 \end{vmatrix} \qquad (464)$$

where the hat indicates a physical component. The last equation may also be given in a more compact form as:

$$[\nabla \times \mathbf{A}]_i = \sum_{k=1}^{3} \frac{\epsilon_{ijk} h_i}{h_1 h_2 h_3} \frac{\partial (h_k \hat{A}_k)}{\partial q^j} \qquad \text{(no sum on } i\text{)} \qquad (465)$$

6.3.5 Laplacian

As seen earlier, for general orthogonal coordinate systems in 3D spaces we have:

$$\sqrt{g} = h_1 h_2 h_3 \qquad g^{ii} = \frac{1}{(h_i)^2} \quad \text{(no sum)} \qquad g^{ij} = 0 \quad (i \neq j) \tag{466}$$

and hence Eq. 454 becomes:

$$\nabla^2 f = \frac{1}{h_1 h_2 h_3} \sum_{i=1}^{3} \frac{\partial}{\partial q^i} \left(\frac{h_1 h_2 h_3}{(h_i)^2} \frac{\partial f}{\partial q^i} \right) \tag{467}$$

which is the Laplacian of a scalar function of position, f, in orthogonal coordinate systems as defined in vector calculus.

6.4 Cylindrical Coordinate System

For cylindrical coordinate systems identified by the coordinates (ρ, ϕ, z), the orthonormal basis vectors are $\mathbf{e}_\rho$, $\mathbf{e}_\phi$ and $\mathbf{e}_z$. Although the given components (i.e. A_ρ, A_ϕ and A_z) are physical components, we do not use hats since the components are suffixed with coordinate symbols (refer to § 3.3). We use for brevity $\mathbf{e}_{\rho\phi}$ as a shorthand notation for the unit dyad $\mathbf{e}_\rho \mathbf{e}_\phi$ and similar notations for the other dyads. We remark that the following expressions for the operators and operations in cylindrical coordinate systems can be obtained from the definition of these operators and operations in general coordinate systems using the metric tensor of the cylindrical system (see § 4.5). They can also be obtained more simply from the corresponding expressions in orthogonal coordinate systems using the scale factors of the cylindrical system in Table 1.

It should be obvious that since ρ, ϕ and z are labels for specific coordinates and not variable indices, the summation convention does not apply to these labels. We note that cylindrical coordinate systems are defined specifically for 3D spaces. We also note that the following operators and operations can be obtained for the 2D plane polar coordinate systems by dropping the z components or terms from the cylindrical form of these operators and operations.

6.4.1 Operators

The nabla operator ∇ in cylindrical coordinate systems is given by:

$$\nabla = \mathbf{e}_\rho \partial_\rho + \mathbf{e}_\phi \frac{1}{\rho} \partial_\phi + \mathbf{e}_z \partial_z \tag{468}$$

while the Laplacian operator is given by:

$$\nabla^2 = \partial_{\rho\rho} + \frac{1}{\rho} \partial_\rho + \frac{1}{\rho^2} \partial_{\phi\phi} + \partial_{zz} \tag{469}$$

where $\partial_{\rho\rho} = \partial_\rho \partial_\rho$ and the notation equally applies to other similar symbols.

6.4.2 Gradient

The gradient of a differentiable scalar f is given by:

$$\nabla f = \mathbf{e}_\rho \partial_\rho f + \mathbf{e}_\phi \frac{1}{\rho} \partial_\phi f + \mathbf{e}_z \partial_z f \tag{470}$$

The gradient of a differentiable vector $\mathbf{A}$ is given by:

$$\begin{aligned}
\nabla \mathbf{A} &= \mathbf{e}_{\rho\rho} A_{\rho,\rho} + \mathbf{e}_{\rho\phi} A_{\phi,\rho} + \mathbf{e}_{\rho z} A_{z,\rho} + \\
&\quad \mathbf{e}_{\phi\rho} \left(\frac{1}{\rho} A_{\rho,\phi} - \frac{A_\phi}{\rho} \right) + \mathbf{e}_{\phi\phi} \left(\frac{1}{\rho} A_{\phi,\phi} + \frac{A_\rho}{\rho} \right) + \mathbf{e}_{\phi z} \frac{1}{\rho} A_{z,\phi} + \\
&\quad \mathbf{e}_{z\rho} A_{\rho,z} + \mathbf{e}_{z\phi} A_{\phi,z} + \mathbf{e}_{zz} A_{z,z}
\end{aligned} \tag{471}$$

where $\mathbf{e}_{\rho\rho}, \mathbf{e}_{\rho\phi}, \cdots, \mathbf{e}_{zz}$ are unit dyads as defined above.

6.4.3 Divergence

The divergence of a differentiable vector $\mathbf{A}$ is given by:

$$\nabla \cdot \mathbf{A} = \frac{1}{\rho} \left[\partial_\rho \left(\rho A_\rho \right) + \partial_\phi A_\phi + \rho \partial_z A_z \right] \tag{472}$$

The divergence of a differentiable rank-2 tensor $\mathbf{A}$ is given by:

$$\begin{aligned}
\nabla \cdot \mathbf{A} &= \mathbf{e}_\rho \left(A_{\rho\rho,\rho} + \frac{A_{\rho\rho} - A_{\phi\phi}}{\rho} + \frac{1}{\rho} A_{\phi\rho,\phi} + A_{z\rho,z} \right) + \\
&\quad \mathbf{e}_\phi \left(A_{\rho\phi,\rho} + \frac{2 A_{\rho\phi}}{\rho} + \frac{1}{\rho} A_{\phi\phi,\phi} + A_{z\phi,z} + \frac{A_{\phi\rho} - A_{\rho\phi}}{\rho} \right) + \\
&\quad \mathbf{e}_z \left(A_{\rho z,\rho} + \frac{A_{\rho z}}{\rho} + \frac{1}{\rho} A_{\phi z,\phi} + A_{zz,z} \right)
\end{aligned} \tag{473}$$

We note that Eq. 472 can be obtained by contracting Eq. 471.

6.4.4 Curl

The curl of a differentiable vector $\mathbf{A}$ is given by:

$$\nabla \times \mathbf{A} = \frac{1}{\rho} \begin{vmatrix} \mathbf{e}_\rho & \rho \mathbf{e}_\phi & \mathbf{e}_z \\ \partial_\rho & \partial_\phi & \partial_z \\ A_\rho & \rho A_\phi & A_z \end{vmatrix} \tag{474}$$

6.4.5 Laplacian

The Laplacian of a differentiable scalar f is given by:

$$\nabla^2 f = \partial_{\rho\rho} f + \frac{1}{\rho} \partial_\rho f + \frac{1}{\rho^2} \partial_{\phi\phi} f + \partial_{zz} f \tag{475}$$

The Laplacian of a differentiable vector $\mathbf{A}$ is given by:

$$\nabla^2 \mathbf{A} = \mathbf{e}_\rho \left[\partial_\rho \left(\frac{1}{\rho} \partial_\rho (\rho A_\rho) \right) + \frac{1}{\rho^2} \partial_{\phi\phi} A_\rho + \partial_{zz} A_\rho - \frac{2}{\rho^2} \partial_\phi A_\phi \right] + \qquad (476)$$

$$\mathbf{e}_\phi \left[\partial_\rho \left(\frac{1}{\rho} \partial_\rho (\rho A_\phi) \right) + \frac{1}{\rho^2} \partial_{\phi\phi} A_\phi + \partial_{zz} A_\phi + \frac{2}{\rho^2} \partial_\phi A_\rho \right] +$$

$$\mathbf{e}_z \left[\frac{1}{\rho} \partial_\rho (\rho \partial_\rho A_z) + \frac{1}{\rho^2} \partial_{\phi\phi} A_z + \partial_{zz} A_z \right]$$

6.5 Spherical Coordinate System

For spherical coordinate systems identified by the coordinates (r, θ, ϕ), the orthonormal basis vectors are $\mathbf{e}_r$, $\mathbf{e}_\theta$ and $\mathbf{e}_\phi$. As in the case of cylindrical coordinates, the components are physical and we do not use hats for the same reason. We use for brevity $\mathbf{e}_{r\theta}$ as a shorthand notation for the dyad $\mathbf{e}_r \mathbf{e}_\theta$ and similar notations for the other unit dyads. As for cylindrical systems, the following expressions for the operators and operations in spherical coordinate systems can be obtained from the corresponding definitions in general coordinate systems using the metric tensor of the spherical system (see § 4.5) or from the corresponding expressions in orthogonal systems using the scale factors of the spherical system in Table 1. Again, the summation convention does not apply to r, θ and ϕ since they are labels and not indices. We also note that spherical coordinate systems are defined specifically for 3D spaces.

6.5.1 Operators

The nabla operator ∇ in spherical coordinate systems is given by:

$$\nabla = \mathbf{e}_r \partial_r + \mathbf{e}_\theta \frac{1}{r} \partial_\theta + \mathbf{e}_\phi \frac{1}{r \sin \theta} \partial_\phi \qquad (477)$$

while the Laplacian operator is given by:

$$\nabla^2 = \partial_{rr} + \frac{2}{r} \partial_r + \frac{1}{r^2} \partial_{\theta\theta} + \frac{\cos \theta}{r^2 \sin \theta} \partial_\theta + \frac{1}{r^2 \sin^2 \theta} \partial_{\phi\phi} \qquad (478)$$

where $\partial_{rr} = \partial_r \partial_r$ and the notation equally applies to other similar symbols.

6.5.2 Gradient

The gradient of a differentiable scalar f in spherical coordinate systems is given by:

$$\nabla f = \mathbf{e}_r \partial_r f + \mathbf{e}_\theta \frac{1}{r} \partial_\theta f + \mathbf{e}_\phi \frac{1}{r \sin \theta} \partial_\phi f \qquad (479)$$

The gradient of a differentiable vector $\mathbf{A}$ is given by:

$$\nabla \mathbf{A} = \mathbf{e}_{rr} A_{r,r} + \mathbf{e}_{r\theta} A_{\theta,r} + \mathbf{e}_{r\phi} A_{\phi,r} + \qquad (480)$$

6.5.3 Divergence

$$\mathbf{e}_{\theta r}\left(\frac{A_{r,\theta}}{r}-\frac{A_\theta}{r}\right)+\mathbf{e}_{\theta\theta}\left(\frac{A_{\theta,\theta}}{r}+\frac{A_r}{r}\right)+\mathbf{e}_{\theta\phi}\frac{A_{\phi,\theta}}{r}+$$

$$\mathbf{e}_{\phi r}\left(\frac{A_{r,\phi}}{r\sin\theta}-\frac{A_\phi}{r}\right)+\mathbf{e}_{\phi\theta}\left(\frac{A_{\theta,\phi}}{r\sin\theta}-\frac{A_\phi\cot\theta}{r}\right)+\mathbf{e}_{\phi\phi}\left(\frac{A_{\phi,\phi}}{r\sin\theta}+\frac{A_r}{r}+\frac{A_\theta\cot\theta}{r}\right)$$

where $\mathbf{e}_{rr}, \mathbf{e}_{r\theta}, \cdots, \mathbf{e}_{\phi\phi}$ are unit dyads as defined above.

6.5.3 Divergence

The divergence of a differentiable vector $\mathbf{A}$ is given by:

$$\nabla\cdot\mathbf{A}=\frac{1}{r^2\sin\theta}\left[\sin\theta\frac{\partial\left(r^2 A_r\right)}{\partial r}+r\frac{\partial\left(\sin\theta A_\theta\right)}{\partial\theta}+r\frac{\partial A_\phi}{\partial\phi}\right] \tag{481}$$

The divergence of a differentiable rank-2 tensor $\mathbf{A}$ is given by:

$$\begin{aligned}\nabla\cdot\mathbf{A}\ =\ & \mathbf{e}_r\left(\frac{\partial_r\left(r^2 A_{rr}\right)}{r^2}+\frac{\partial_\theta\left(A_{\theta r}\sin\theta\right)}{r\sin\theta}+\frac{\partial_\phi A_{\phi r}}{r\sin\theta}-\frac{A_{\theta\theta}+A_{\phi\phi}}{r}\right)+ \\ & \mathbf{e}_\theta\left(\frac{\partial_r\left(r^3 A_{r\theta}\right)}{r^3}+\frac{\partial_\theta\left(A_{\theta\theta}\sin\theta\right)}{r\sin\theta}+\frac{\partial_\phi A_{\phi\theta}}{r\sin\theta}+\frac{A_{\theta r}-A_{r\theta}-A_{\phi\phi}\cot\theta}{r}\right)+ \\ & \mathbf{e}_\phi\left(\frac{\partial_r\left(r^3 A_{r\phi}\right)}{r^3}+\frac{\partial_\theta\left(A_{\theta\phi}\sin\theta\right)}{r\sin\theta}+\frac{\partial_\phi A_{\phi\phi}}{r\sin\theta}+\frac{A_{\phi r}-A_{r\phi}+A_{\phi\theta}\cot\theta}{r}\right)\end{aligned} \tag{482}$$

We note that Eq. 481 can be obtained by contracting Eq. 480.

6.5.4 Curl

The curl of a differentiable vector $\mathbf{A}$ is given by:

$$\nabla\times\mathbf{A}=\frac{1}{r^2\sin\theta}\begin{vmatrix}\mathbf{e}_r & r\mathbf{e}_\theta & r\sin\theta\mathbf{e}_\phi \\ \partial_r & \partial_\theta & \partial_\phi \\ A_r & rA_\theta & r\sin\theta A_\phi\end{vmatrix} \tag{483}$$

6.5.5 Laplacian

The Laplacian of a differentiable scalar f is given by:

$$\nabla^2 f=\partial_{rr}f+\frac{2}{r}\partial_r f+\frac{1}{r^2}\partial_{\theta\theta}f+\frac{\cos\theta}{r^2\sin\theta}\partial_\theta f+\frac{1}{r^2\sin^2\theta}\partial_{\phi\phi}f \tag{484}$$

The Laplacian of a differentiable vector $\mathbf{A}$ is given by:

$$\begin{aligned}\nabla^2\mathbf{A}\ =\ & \mathbf{e}_r\left[\partial_r\left(\frac{\partial_r\left(r^2 A_r\right)}{r^2}\right)+\frac{\partial_\theta\left(\sin\theta\partial_\theta A_r\right)}{r^2\sin\theta}+\frac{\partial_{\phi\phi}A_r}{r^2\sin^2\theta}-\frac{2\partial_\theta\left(A_\theta\sin\theta\right)}{r^2\sin\theta}-\frac{2\partial_\phi A_\phi}{r^2\sin\theta}\right]+ \\ & \mathbf{e}_\theta\left[\frac{\partial_r\left(r^2\partial_r A_\theta\right)}{r^2}+\frac{1}{r^2}\partial_\theta\left(\frac{\partial_\theta\left(A_\theta\sin\theta\right)}{\sin\theta}\right)+\frac{\partial_{\phi\phi}A_\theta}{r^2\sin^2\theta}+\frac{2\partial_\theta A_r}{r^2}-\frac{2\cot\theta}{r^2\sin\theta}\partial_\phi A_\phi\right]+ \\ & \mathbf{e}_\phi\left[\frac{\partial_r\left(r^2\partial_r A_\phi\right)}{r^2}+\frac{1}{r^2}\partial_\theta\left(\frac{\partial_\theta\left(A_\phi\sin\theta\right)}{\sin\theta}\right)+\frac{\partial_{\phi\phi}A_\phi}{r^2\sin^2\theta}+\frac{2\partial_\phi A_r}{r^2\sin\theta}+\frac{2\cot\theta}{r^2\sin\theta}\partial_\phi A_\theta\right]\end{aligned} \tag{485}$$

6.6 Exercises and Revision

6.1 Describe briefly the nabla based differential operators and operations considering the interaction of the nabla operator with the tensors which are acted upon by this operator.

6.2 What are the advantages and disadvantages of using the coordinates as suffixes for labeling the operators, basis vectors and tensor components in cylindrical and spherical systems instead of indexed general coordinates? What are the advantages and disadvantages of the opposite?

6.3 "The differentiation of a tensor increases its rank by one, by introducing an extra covariant index, unless it implies a contraction in which case it reduces the rank by one". Justify this statement giving common examples from vector and tensor calculus.

6.4 Write the following subsidiary nabla based operators in tensor notation: $\mathbf{A} \cdot \nabla$ and $\mathbf{A} \times \nabla$. Is this notation consistent with the notation of dot and cross product of vectors?

6.5 Why in general we have: $\mathbf{A} \cdot \nabla \neq \nabla \cdot \mathbf{A}$ and $\mathbf{A} \times \nabla \neq \nabla \times \mathbf{A}$?

6.6 Define the nabla vector operator and the Laplacian scalar operator in Cartesian coordinate systems using tensor notation.

6.7 Find the gradient of the following vector field in a Cartesian coordinate system: $\mathbf{A} = (x, 2x^2, \pi)$.

6.8 Define the divergence of a differentiable vector descriptively and mathematically assuming a Cartesian coordinate system.

6.9 What is the divergence of the following vector field in Cartesian coordinates: $\mathbf{A} = (2z, y^3, e^x)$?

6.10 Write symbolically, using tensor notation, the following two forms of the divergence of a rank-2 tensor field $\mathbf{A}$ in Cartesian coordinates: $\nabla \cdot \mathbf{A}$ and $\nabla \cdot \mathbf{A}^T$.

6.11 Define the curl $\nabla \times \mathbf{A}$ in Cartesian coordinates using tensor notation where (a) $\mathbf{A}$ is a rank-1 tensor and (b) $\mathbf{A}$ is a rank-2 tensor (note the two possibilities in the last case).

6.12 What is the curl of the following vector field assuming a Cartesian coordinate system: $\mathbf{A} = (5e^{2x}, \pi xy, z^2)$?

6.13 Find the Laplacian of the following vector field in Cartesian coordinates: $\mathbf{A} = (x^2 y, 2y \sin z, \pi z e^{\cosh x})$.

6.14 Define the nabla operator and the Laplacian operator in general coordinate systems using tensor notation.

6.15 Obtain an expression for the gradient of a covariant vector $\mathbf{A} = A_i \mathbf{E}^i$ in general coordinates justifying each step in your derivation. Repeat the question with a contravariant vector $\mathbf{A} = A^i \mathbf{E}_i$.

6.16 Repeat question 6.15 with a rank-2 mixed tensor $\mathbf{A} = A^i_{\ j} \mathbf{E}_i \mathbf{E}^j$.

6.17 Define, in tensor language, the contravariant form of the gradient of a scalar field.

6.18 Define the divergence of a differentiable vector descriptively and mathematically assuming a general coordinate system.

6.19 Derive the following expression for the divergence of a contravariant vector $\mathbf{A}$ in

6.6 Exercises and Revision

general coordinates: $\nabla \cdot \mathbf{A} = A^i_{;i}$.

6.20 Verify the following formula for the divergence of a contravariant vector $\mathbf{A}$ in general coordinates: $\nabla \cdot \mathbf{A} = \frac{1}{\sqrt{g}} \partial_i \left(\sqrt{g} A^i \right)$. Repeat the question with the formula: $\nabla \cdot \mathbf{A} = g^{ji} A_{j;i}$ where $\mathbf{A}$ is a covariant vector.

6.21 Repeat question 6.20 with the formula: $\nabla \cdot \mathbf{A} = \mathbf{E}_k A^{ik}_{;i}$ where $\mathbf{A}$ is a rank-2 contravariant tensor.

6.22 Prove that the divergence of a contravariant vector is a scalar (i.e. rank-0 tensor) by showing that it is invariant under coordinate transformations.

6.23 Derive, from the first principles, the following formula for the curl of a covariant vector field $\mathbf{A}$ in general coordinates: $[\nabla \times \mathbf{A}]^k = \frac{\epsilon^{ijk}}{\sqrt{g}} \left(\partial_i A_j - \Gamma^l_{ji} A_l \right)$.

6.24 Show that the formula in exercise 6.23 will reduce to $[\nabla \times \mathbf{A}]^k = \frac{\epsilon^{ijk}}{\sqrt{g}} \partial_i A_j$ due to the symmetry of the Christoffel symbols in their lower indices.

6.25 Derive, from the first principles, the following expression for the Laplacian of a scalar field f in general coordinates: $\nabla^2 f = \frac{1}{\sqrt{g}} \partial_i \left(\sqrt{g} g^{ij} \partial_j f \right)$.

6.26 Why the basic definition of the Laplacian of a scalar field f in general coordinates as $\nabla^2 f = \text{div} (\text{grad } f)$ cannot be used as it is to develop a formula before raising the index of the gradient?

6.27 Define, in tensor language, the nabla operator and the Laplacian operator assuming an orthogonal coordinate system of a 3D space.

6.28 Using the expression of the divergence of a vector field in general coordinates, obtain an expression for the divergence in orthogonal coordinates.

6.29 Define the curl of a vector field $\mathbf{A}$ in orthogonal coordinates of a 3D space using determinantal form and tensor notation form.

6.30 Using the expression of the Laplacian of a scalar field in general coordinates, derive an expression for the Laplacian in orthogonal coordinates of a 3D space.

6.31 Why the components of tensors in cylindrical and spherical coordinates are physical?

6.32 Define the nabla and Laplacian operators in cylindrical coordinates.

6.33 Use the definition of the gradient of a scalar field f in orthogonal coordinates and the scale factors of Table 1 to obtain an expression for the gradient in cylindrical coordinates.

6.34 Use the definition of the divergence of a vector field $\mathbf{A}$ in orthogonal coordinates and the scale factors of Table 1 to obtain an expression for the divergence in cylindrical coordinates.

6.35 Write the determinantal form of the curl of a vector field $\mathbf{A}$ in cylindrical coordinates.

6.36 Use the definition of the Laplacian of a scalar field f in orthogonal coordinates and the scale factors of Table 1 to obtain an expression for the Laplacian in cylindrical coordinates.

6.37 A scalar field in cylindrical coordinates is given by: $f(\rho, \phi, z) = \rho$. What are the gradient and Laplacian of this field?

6.38 A vector field in cylindrical coordinates is given by: $\mathbf{A}(\rho, \phi, z) = (3z, \pi \phi^2, z^2 \cos \rho)$. What are the divergence and curl of this field?

6.39 Repeat exercise 6.33 with spherical coordinates.

6.40 Repeat exercise 6.34 with spherical coordinates.
6.41 Repeat exercise 6.35 with spherical coordinates.
6.42 Repeat exercise 6.36 with spherical coordinates.
6.43 A scalar field in spherical coordinates is given by: $f(r,\theta,\phi) = r^2 + \theta$. What are the gradient and Laplacian of this field?
6.44 A vector field in spherical coordinates is given by: $\mathbf{A}(r,\theta,\phi) = (e^r, 5\sin\phi, \ln\theta)$. What are the divergence and curl of this field?

Chapter 7
Tensors in Application

In this chapter, we conduct a preliminary investigation about some tensors and tensor notation and techniques which are commonly used in the mathematical and physical applications of tensor calculus. The chapter is made of three sections dedicated to tensors in mathematics, geometry and science. The mathematics part (see § 7.1) comes from tensor applications in linear algebra and vector calculus, while most of the materials in the science part (see § 7.3) come from applications related to fluid and continuum mechanics. In the geometry part (see § 7.2) a few prominent tensors and tensor identities of wide applications in differential geometry and related scientific fields are examined. We should also refer the reader to the previous chapter (see § 6) as an example of tensor applications in mathematics and science since the materials in that chapter are partly based on employing tensor notation and techniques.

We note that all the aforementioned disciplines, where the materials about tensors come from, are intimately linked to tensor calculus since large parts of this subject were developed within those disciplines. Also, the materials presented in this chapter about tensors and tensor techniques are used as vital building blocks and tools in a number of important mathematical and physical theories. However, we would like to insist that although these materials provide a very useful glimpse, they are just partially representative examples of tensor applications. Our objective is to have more familiarity with some prominent tensors and tensor techniques and hence they are not meant to provide a comprehensive view. We should also indicate that some tensors, especially those in the science section, are defined differently in different disciplines and hence the given definitions and descriptions may not be thorough or universal.

7.1 Tensors in Mathematics

In this section, we provide a sample of common definitions related to basic concepts and operations in matrix and vector algebra using the language of tensor calculus. Common identities in vector calculus as well as the integral theorems of Gauss and Stokes and some important scalar invariants are also examined in this section from this perspective. Finally, we provide a rather extensive set of examples about the use of tensor language and techniques in proving mathematical identities where these identities are gathered from vector calculus. For simplicity, clarity and wide applicability we employ a Cartesian approach in the tensor formulations of this section.

7.1.1 Common Definitions in Tensor Notation

The trace of a matrix $\mathbf{A}$ representing a rank-2 tensor in an nD space is given by:

$$\mathrm{tr}\left(\mathbf{A}\right) = A_{ii} \qquad (i = 1, \ldots, n) \tag{486}$$

For a 3×3 matrix representing a rank-2 tensor in a 3D space, the determinant is given by:

$$\det\left(\mathbf{A}\right) = \begin{vmatrix} A_{11} & A_{12} & A_{13} \\ A_{21} & A_{22} & A_{23} \\ A_{31} & A_{32} & A_{33} \end{vmatrix} = \epsilon_{ijk} A_{1i} A_{2j} A_{3k} = \epsilon_{ijk} A_{i1} A_{j2} A_{k3} \tag{487}$$

where the last two equalities represent the expansion of the determinant by row and by column. Alternatively, the determinant of a 3×3 matrix can be given by:

$$\det\left(\mathbf{A}\right) = \frac{1}{3!} \epsilon_{ijk} \epsilon_{lmn} A_{il} A_{jm} A_{kn} \tag{488}$$

More generally, for an $n \times n$ matrix representing a rank-2 tensor in an nD space, the determinant is given by:

$$\begin{aligned} \det\left(\mathbf{A}\right) &= \epsilon_{i_1 \cdots i_n} A_{1 i_1} \ldots A_{n i_n} \\ &= \epsilon_{i_1 \cdots i_n} A_{i_1 1} \ldots A_{i_n n} \\ &= \frac{1}{n!} \epsilon_{i_1 \cdots i_n} \epsilon_{j_1 \cdots j_n} A_{i_1 j_1} \ldots A_{i_n j_n} \end{aligned} \tag{489}$$

The inverse of a matrix $\mathbf{A}$ representing a rank-2 tensor is given by:

$$\left[\mathbf{A}^{-1}\right]_{ij} = \frac{1}{2 \det\left(\mathbf{A}\right)} \epsilon_{jmn} \epsilon_{ipq} A_{mp} A_{nq} \tag{490}$$

The multiplication of a matrix $\mathbf{A}$ by a vector $\mathbf{b}$, as defined in linear algebra, is given by:

$$[\mathbf{A}\mathbf{b}]_i = A_{ij} b_j \tag{491}$$

It should be remarked that we are using matrix notation in the writing of $\mathbf{Ab}$. According to the symbolic notation of tensors, the multiplication operation should be denoted by a dot between the symbols of the tensor and the vector, i.e. $\mathbf{A}\cdot\mathbf{b}$, since matrix multiplication in matrix algebra is equivalent to an inner product operation in tensor algebra.

Similarly, the multiplication of two compatible matrices $\mathbf{A}$ and $\mathbf{B}$, as defined in linear algebra, is given by:

$$[\mathbf{AB}]_{ik} = A_{ij} B_{jk} \tag{492}$$

Again, we are using here matrix notation in the writing of $\mathbf{AB}$; otherwise a dot should be inserted between the symbols of the two matrices.

The dot product of two vectors of the same dimension is given by:

$$\mathbf{A} \cdot \mathbf{B} = \delta_{ij} A_i B_j = A_i B_i \tag{493}$$

7.1.2 Scalar Invariants of Tensors

Similarly, the cross product of two vectors in a 3D space is given by:

$$[\mathbf{A} \times \mathbf{B}]_i = \epsilon_{ijk} A_j B_k \tag{494}$$

The scalar triple product of three vectors in a 3D space is given by:

$$\mathbf{A} \cdot (\mathbf{B} \times \mathbf{C}) = \begin{vmatrix} A_1 & A_2 & A_3 \\ B_1 & B_2 & B_3 \\ C_1 & C_2 & C_3 \end{vmatrix} = \epsilon_{ijk} A_i B_j C_k \tag{495}$$

while the vector triple product of three vectors in a 3D space is given by:

$$[\mathbf{A} \times (\mathbf{B} \times \mathbf{C})]_i = \epsilon_{ijk} \epsilon_{klm} A_j B_l C_m \tag{496}$$

The expression of the other principal form of the vector triple product, i.e. $(\mathbf{A} \times \mathbf{B}) \times \mathbf{C}$, can be obtained from the above form by changing the order of the factors in the external cross product and reversing the sign. Other operations, like relabeling the indices and exchanging some of the indices of the epsilons with a shift in sign, can then follow to obtain a more organized form.

7.1.2 Scalar Invariants of Tensors

In the following, we list and write in tensor notation a number of invariants of low rank tensors which have special importance due to their widespread applications in vector and tensor calculus. All these invariants are scalars.

The value of a scalar (rank-0 tensor), which consists of a magnitude and a sign, is invariant under coordinate transformations. An invariant of a vector (rank-1 tensor) under coordinate transformations is its magnitude, i.e. length.[29] The main three independent scalar invariants of a rank-2 tensor $\mathbf{A}$ are:

$$\begin{align} I &= \text{tr}(\mathbf{A}) = A_{ii} \tag{497} \\ II &= \text{tr}(\mathbf{A}^2) = A_{ij} A_{ji} \tag{498} \\ III &= \text{tr}(\mathbf{A}^3) = A_{ij} A_{jk} A_{ki} \tag{499} \end{align}$$

Different forms of the three invariants of a rank-2 tensor $\mathbf{A}$, which are also widely used, are the following:

$$\begin{align} I_1 &= I = A_{ii} \tag{500} \\ I_2 &= \frac{1}{2}\left(I^2 - II\right) = \frac{1}{2}\left(A_{ii} A_{jj} - A_{ij} A_{ji}\right) \tag{501} \\ I_3 &= \det(\mathbf{A}) = \frac{1}{3!}\left(I^3 - 3I\,II + 2III\right) = \frac{1}{3!} \epsilon_{ijk} \epsilon_{pqr} A_{ip} A_{jq} A_{kr} \tag{502} \end{align}$$

---

[29] The direction is also invariant but it is not a scalar! In fact the magnitude alone is invariant under coordinate transformations even for pseudo vectors because it is a true scalar.

where I, II and III are given by Eqs. 497-499. As the invariants I_1, I_2 and I_3 are defined in terms of the invariants I, II and III according to Eqs. 500-502, the invariants I, II and III can also be defined in terms of the invariants I_1, I_2 and I_3, that is:

$$I = I_1 \tag{503}$$
$$II = I_1^2 - 2I_2 \tag{504}$$
$$III = I_1^3 - 3I_1 I_2 + 3I_3 \tag{505}$$

Since the determinant of a matrix representing a rank-2 tensor is invariant (see Eq. 502), then if the determinant vanishes in one coordinate system, it will vanish in all transformed coordinate systems, and if not it will not (also refer to § 3.1.4). Consequently, if a rank-2 tensor is invertible in a particular coordinate system, it will be invertible in all coordinate systems, and if not it will not.

The following are ten common scalar invariants that are jointly formed between two rank-2 tensors $\mathbf{A}$ and $\mathbf{B}$: $\text{tr}(\mathbf{A})$, $\text{tr}(\mathbf{B})$, $\text{tr}(\mathbf{A}^2)$, $\text{tr}(\mathbf{B}^2)$, $\text{tr}(\mathbf{A}^3)$, $\text{tr}(\mathbf{B}^3)$, $\text{tr}(\mathbf{A} \cdot \mathbf{B})$, $\text{tr}(\mathbf{A}^2 \cdot \mathbf{B})$, $\text{tr}(\mathbf{A} \cdot \mathbf{B}^2)$ and $\text{tr}(\mathbf{A}^2 \cdot \mathbf{B}^2)$. As seen, all these are traces.

7.1.3 Common Identities in Vector and Tensor Notation

In this subsection, we present some of the widely used identities of vector calculus using the traditional vector notation as well as its equivalent tensor notation. In the following equations, f and h are differentiable scalar fields; $\mathbf{A}$, $\mathbf{B}$, $\mathbf{C}$ and $\mathbf{D}$ are differentiable vector fields; and $\mathbf{r} = x_i \mathbf{e}_i$ is the position vector.

$$\nabla \cdot \mathbf{r} = n \quad \text{(vector notation)} \tag{506}$$
$$\partial_i x_i = n \quad \text{(tensor notation)} \tag{507}$$

where n is the space dimension.

$$\nabla \times \mathbf{r} = \mathbf{0} \tag{508}$$
$$\epsilon_{ijk} \partial_j x_k = 0 \tag{509}$$

$$\nabla (\mathbf{a} \cdot \mathbf{r}) = \mathbf{a} \tag{510}$$
$$\partial_i (a_j x_j) = a_i \tag{511}$$

where $\mathbf{a}$ is a constant vector.

$$\nabla \cdot (\nabla f) = \nabla^2 f \tag{512}$$
$$\partial_i (\partial_i f) = \partial_{ii} f \tag{513}$$

$$\nabla \cdot (\nabla \times \mathbf{A}) = 0 \tag{514}$$
$$\epsilon_{ijk} \partial_i \partial_j A_k = 0 \tag{515}$$

7.1.3 Common Identities in Vector and Tensor Notation

$$\nabla \times (\nabla f) = \mathbf{0} \tag{516}$$
$$\epsilon_{ijk}\partial_j\partial_k f = 0 \tag{517}$$

$$\nabla(fh) = f\nabla h + h\nabla f \tag{518}$$
$$\partial_i(fh) = f\partial_i h + h\partial_i f \tag{519}$$

$$\nabla \cdot (f\mathbf{A}) = f\nabla \cdot \mathbf{A} + \mathbf{A} \cdot \nabla f \tag{520}$$
$$\partial_i(fA_i) = f\partial_i A_i + A_i\partial_i f \tag{521}$$

$$\nabla \times (f\mathbf{A}) = f\nabla \times \mathbf{A} + \nabla f \times \mathbf{A} \tag{522}$$
$$\epsilon_{ijk}\partial_j(fA_k) = f\epsilon_{ijk}\partial_j A_k + \epsilon_{ijk}(\partial_j f)A_k \tag{523}$$

$$\mathbf{A} \cdot (\mathbf{B} \times \mathbf{C}) = \mathbf{C} \cdot (\mathbf{A} \times \mathbf{B}) = \mathbf{B} \cdot (\mathbf{C} \times \mathbf{A}) \tag{524}$$
$$\epsilon_{ijk}A_iB_jC_k = \epsilon_{kij}C_kA_iB_j = \epsilon_{jki}B_jC_kA_i \tag{525}$$

$$\mathbf{A} \times (\mathbf{B} \times \mathbf{C}) = \mathbf{B}(\mathbf{A} \cdot \mathbf{C}) - \mathbf{C}(\mathbf{A} \cdot \mathbf{B}) \tag{526}$$
$$\epsilon_{ijk}A_j\epsilon_{klm}B_lC_m = B_i(A_mC_m) - C_i(A_lB_l) \tag{527}$$

$$\mathbf{A} \times (\nabla \times \mathbf{B}) = (\nabla \mathbf{B}) \cdot \mathbf{A} - \mathbf{A} \cdot \nabla \mathbf{B} \tag{528}$$
$$\epsilon_{ijk}\epsilon_{klm}A_j\partial_l B_m = (\partial_i B_m)A_m - A_l(\partial_l B_i) \tag{529}$$

$$\nabla \times (\nabla \times \mathbf{A}) = \nabla(\nabla \cdot \mathbf{A}) - \nabla^2 \mathbf{A} \tag{530}$$
$$\epsilon_{ijk}\epsilon_{klm}\partial_j\partial_l A_m = \partial_i(\partial_m A_m) - \partial_{ll} A_i \tag{531}$$

$$\nabla(\mathbf{A} \cdot \mathbf{B}) = \mathbf{A} \times (\nabla \times \mathbf{B}) + \mathbf{B} \times (\nabla \times \mathbf{A}) + (\mathbf{A} \cdot \nabla)\mathbf{B} + (\mathbf{B} \cdot \nabla)\mathbf{A} \tag{532}$$
$$\partial_i(A_m B_m) = \epsilon_{ijk}A_j(\epsilon_{klm}\partial_l B_m) + \epsilon_{ijk}B_j(\epsilon_{klm}\partial_l A_m) + (A_l\partial_l)B_i + (B_l\partial_l)A_i \tag{533}$$

$$\nabla \cdot (\mathbf{A} \times \mathbf{B}) = \mathbf{B} \cdot (\nabla \times \mathbf{A}) - \mathbf{A} \cdot (\nabla \times \mathbf{B}) \tag{534}$$
$$\partial_i(\epsilon_{ijk}A_jB_k) = B_k(\epsilon_{kij}\partial_i A_j) - A_j(\epsilon_{jik}\partial_i B_k) \tag{535}$$

$$\nabla \times (\mathbf{A} \times \mathbf{B}) = (\mathbf{B} \cdot \nabla)\mathbf{A} + (\nabla \cdot \mathbf{B})\mathbf{A} - (\nabla \cdot \mathbf{A})\mathbf{B} - (\mathbf{A} \cdot \nabla)\mathbf{B} \tag{536}$$
$$\epsilon_{ijk}\epsilon_{klm}\partial_j(A_lB_m) = (B_m\partial_m)A_i + (\partial_m B_m)A_i - (\partial_j A_j)B_i - (A_j\partial_j)B_i \tag{537}$$

$$(\mathbf{A} \times \mathbf{B}) \cdot (\mathbf{C} \times \mathbf{D}) = \begin{vmatrix} \mathbf{A} \cdot \mathbf{C} & \mathbf{A} \cdot \mathbf{D} \\ \mathbf{B} \cdot \mathbf{C} & \mathbf{B} \cdot \mathbf{D} \end{vmatrix} \tag{538}$$
$$\epsilon_{ijk}A_jB_k\epsilon_{ilm}C_lD_m = (A_lC_l)(B_mD_m) - (A_mD_m)(B_lC_l) \tag{539}$$

$$(\mathbf{A} \times \mathbf{B}) \times (\mathbf{C} \times \mathbf{D}) = [\mathbf{D} \cdot (\mathbf{A} \times \mathbf{B})]\mathbf{C} - [\mathbf{C} \cdot (\mathbf{A} \times \mathbf{B})]\mathbf{D} \tag{540}$$

$$\epsilon_{ijk}\epsilon_{jmn}A_m B_n \epsilon_{kpq}C_p D_q = (\epsilon_{qmn}D_q A_m B_n)C_i - (\epsilon_{pmn}C_p A_m B_n)D_i \tag{541}$$

In vector and tensor notations, the condition for a vector field $\mathbf{A}$ to be solenoidal is given by:

$$\nabla \cdot \mathbf{A} = 0 \tag{542}$$
$$\partial_i A_i = 0 \tag{543}$$

In vector and tensor notations, the condition for a vector field $\mathbf{A}$ to be irrotational is given by:

$$\nabla \times \mathbf{A} = 0 \tag{544}$$
$$\epsilon_{ijk}\partial_j A_k = 0 \tag{545}$$

7.1.4 Integral Theorems in Tensor Notation

The divergence theorem for a differentiable vector field $\mathbf{A}$ in vector and tensor notations is given by:

$$\iiint_\Omega \nabla \cdot \mathbf{A}\, d\tau = \iint_S \mathbf{A} \cdot \mathbf{n}\, d\sigma \tag{546}$$

$$\int_\Omega \partial_i A_i d\tau = \int_S A_i n_i d\sigma \tag{547}$$

where Ω is a bounded region in an nD space enclosed by a generalized surface S, $d\tau$ and $d\sigma$ are generalized volume and area differentials, $\mathbf{n}$ and n_i are the unit vector normal to the surface and its ith component, and the index i ranges over $1,\ldots,n$.

Similarly, the divergence theorem for a differentiable rank-2 tensor field $\mathbf{A}$ in tensor notation for the first index is given by:

$$\int_\Omega \partial_i A_{il} d\tau = \int_S A_{il} n_i d\sigma \tag{548}$$

while the divergence theorem for differentiable tensor fields of higher rank $\mathbf{A}$ in tensor notation for the index k is given by:

$$\int_\Omega \partial_k A_{ij\ldots k\ldots m} d\tau = \int_S A_{ij\ldots k\ldots m} n_k d\sigma \tag{549}$$

Stokes theorem for a differentiable vector field $\mathbf{A}$ in vector and tensor notations is given by:

$$\iint_S (\nabla \times \mathbf{A}) \cdot \mathbf{n}\, d\sigma = \int_C \mathbf{A} \cdot d\mathbf{r} \tag{550}$$

$$\int_S \epsilon_{ijk}\partial_j A_k n_i d\sigma = \int_C A_i dx_i \tag{551}$$

where C stands for the perimeter of the surface S, and $d\mathbf{r}$ is a differential of the position vector which is tangent to the perimeter while the other symbols are as defined above.

Similarly, Stokes theorem for a differentiable rank-2 tensor field $\mathbf{A}$ in tensor notation for the first index is given by:

$$\int_S \epsilon_{ijk}\partial_j A_{kl} n_i d\sigma = \int_C A_{il} dx_i \tag{552}$$

while Stokes theorem for differentiable tensor fields of higher rank $\mathbf{A}$ in tensor notation for the index k is given by:

$$\int_S \epsilon_{ijk}\partial_j A_{lm...k...n} n_i d\sigma = \int_C A_{lm...k...n} dx_k \tag{553}$$

7.1.5 Examples of Using Tensor Techniques to Prove Identities

In this subsection, we provide some examples for using tensor techniques to prove vector and tensor identities where comments are added next to each step to explain and justify. These examples, which are based on the identities given in § 7.1.3, demonstrate the elegance, efficiency and clarity of the methods and notation of tensor calculus. We note that in Cartesian coordinate systems, some tensor equations in general coordinates change their form and hence in the following we give the corresponding Cartesian form of the equations which are not given previously in this form since we use the Cartesian form in these proofs. The added comments inside the parentheses refer to the corresponding equations in general coordinates:

$$\epsilon_{ijk}\epsilon_{lmk} = \delta_{il}\delta_{jm} - \delta_{im}\delta_{jl} \qquad \text{(Eq. 167)} \tag{554}$$

$$\delta_{ij}A_j = A_i \qquad \text{(Eq. 172)} \tag{555}$$

$$\frac{\partial x_i}{\partial x_j} = \partial_j x_i = x_{i,j} = \delta_{ij} \qquad \text{(Eq. 175)} \tag{556}$$

$$\partial_i x_i = \delta_{ii} = n \qquad \text{(Eq. 176)} \tag{557}$$

- $\nabla \cdot \mathbf{r} = n$:

$$\nabla \cdot \mathbf{r} = \partial_i x_i \qquad \text{(Eq. 419)}$$
$$= \delta_{ii} \qquad \text{(Eq. 557)}$$
$$= n \qquad \text{(Eq. 557)}$$

- $\nabla \times \mathbf{r} = \mathbf{0}$:

$$[\nabla \times \mathbf{r}]_i = \epsilon_{ijk}\partial_j x_k \qquad \text{(Eq. 423)}$$
$$= \epsilon_{ijk}\delta_{kj} \qquad \text{(Eq. 556)}$$
$$= \epsilon_{ijj} \qquad \text{(Eq. 555)}$$
$$= 0 \qquad \text{(Eq. 155)}$$

Since i is a free index, the identity is proved for all components.

7.1.5 Examples of Using Tensor Techniques to Prove Identities

- $\nabla (\mathbf{a} \cdot \mathbf{r}) = \mathbf{a}$:

$$\begin{aligned}
\left[\nabla (\mathbf{a} \cdot \mathbf{r})\right]_i &= \partial_i (a_j x_j) && \text{(Eqs. 417 \& 493)} \\
&= a_j \partial_i x_j + x_j \partial_i a_j && \text{(product rule)} \\
&= a_j \partial_i x_j && (a_j \text{ is constant}) \\
&= a_j \delta_{ji} && \text{(Eq. 556)} \\
&= a_i && \text{(Eq. 555)} \\
&= [\mathbf{a}]_i && \text{(definition of index)}
\end{aligned}$$

Since i is a free index, the identity is proved for all components.

- $\nabla \cdot (\nabla f) = \nabla^2 f$:

$$\begin{aligned}
\nabla \cdot (\nabla f) &= \partial_i [\nabla f]_i && \text{(Eq. 419)} \\
&= \partial_i (\partial_i f) && \text{(Eq. 417)} \\
&= \partial_i \partial_i f && \text{(rules of differentiation)} \\
&= \partial_{ii} f && \text{(definition of 2nd derivative)} \\
&= \nabla^2 f && \text{(Eq. 425)}
\end{aligned}$$

- $\nabla \cdot (\nabla \times \mathbf{A}) = 0$:

$$\begin{aligned}
\nabla \cdot (\nabla \times \mathbf{A}) &= \partial_i [\nabla \times \mathbf{A}]_i && \text{(Eq. 419)} \\
&= \partial_i (\epsilon_{ijk} \partial_j A_k) && \text{(Eq. 423)} \\
&= \epsilon_{ijk} \partial_i \partial_j A_k && (\partial \text{ not acting on } \epsilon) \\
&= \epsilon_{ijk} \partial_j \partial_i A_k && \text{(continuity condition)} \\
&= -\epsilon_{jik} \partial_j \partial_i A_k && \text{(Eq. 180)} \\
&= -\epsilon_{ijk} \partial_i \partial_j A_k && \text{(relabeling dummy indices } i \text{ and } j\text{)} \\
&= 0 && \text{(since } \epsilon_{ijk} \partial_i \partial_j A_k = -\epsilon_{ijk} \partial_i \partial_j A_k\text{)}
\end{aligned}$$

This can also be concluded from line three by arguing that: since by the continuity condition ∂_i and ∂_j can change their order with no change in the value of the term while a corresponding change of the order of i and j in ϵ_{ijk} results in a sign change, we see that each term in the sum has its own negative and hence the terms add up to zero.

- $\nabla \times (\nabla f) = \mathbf{0}$:

$$\begin{aligned}
\left[\nabla \times (\nabla f)\right]_i &= \epsilon_{ijk} \partial_j [\nabla f]_k && \text{(Eq. 423)} \\
&= \epsilon_{ijk} \partial_j (\partial_k f) && \text{(Eq. 417)} \\
&= \epsilon_{ijk} \partial_j \partial_k f && \text{(rules of differentiation)} \\
&= \epsilon_{ijk} \partial_k \partial_j f && \text{(continuity condition)} \\
&= -\epsilon_{ikj} \partial_k \partial_j f && \text{(Eq. 180)} \\
&= -\epsilon_{ijk} \partial_j \partial_k f && \text{(relabeling dummy indices } j \text{ and } k\text{)} \\
&= 0 && \text{(since } \epsilon_{ijk} \partial_j \partial_k f = -\epsilon_{ijk} \partial_j \partial_k f\text{)}
\end{aligned}$$

This can also be concluded from line three by a similar argument to the one given in the previous point. Because $[\nabla \times (\nabla f)]_i$ is an arbitrary component, then each component is zero.

7.1.5 Examples of Using Tensor Techniques to Prove Identities

- $\nabla(fh) = f\nabla h + h\nabla f$:

$$[\nabla(fh)]_i = \partial_i(fh) \qquad \text{(Eq. 417)}$$
$$= f\partial_i h + h\partial_i f \qquad \text{(product rule)}$$
$$= [f\nabla h]_i + [h\nabla f]_i \qquad \text{(Eq. 417)}$$
$$= [f\nabla h + h\nabla f]_i \qquad \text{(Eq. 24)}$$

Because i is a free index, the identity is proved for all components.

- $\nabla \cdot (f\mathbf{A}) = f\nabla \cdot \mathbf{A} + \mathbf{A} \cdot \nabla f$:

$$\nabla \cdot (f\mathbf{A}) = \partial_i [f\mathbf{A}]_i \qquad \text{(Eq. 419)}$$
$$= \partial_i (fA_i) \qquad \text{(definition of index)}$$
$$= f\partial_i A_i + A_i \partial_i f \qquad \text{(product rule)}$$
$$= f\nabla \cdot \mathbf{A} + \mathbf{A} \cdot \nabla f \qquad \text{(Eqs. 419 \& 413)}$$

- $\nabla \times (f\mathbf{A}) = f\nabla \times \mathbf{A} + \nabla f \times \mathbf{A}$:

$$[\nabla \times (f\mathbf{A})]_i = \epsilon_{ijk}\partial_j [f\mathbf{A}]_k \qquad \text{(Eq. 423)}$$
$$= \epsilon_{ijk}\partial_j (fA_k) \qquad \text{(definition of index)}$$
$$= f\epsilon_{ijk}\partial_j A_k + \epsilon_{ijk}(\partial_j f)A_k \qquad \text{(product rule \& commutativity)}$$
$$= f\epsilon_{ijk}\partial_j A_k + \epsilon_{ijk}[\nabla f]_j A_k \qquad \text{(Eq. 417)}$$
$$= [f\nabla \times \mathbf{A}]_i + [\nabla f \times \mathbf{A}]_i \qquad \text{(Eqs. 423 \& 494)}$$
$$= [f\nabla \times \mathbf{A} + \nabla f \times \mathbf{A}]_i \qquad \text{(Eq. 24)}$$

Because i is a free index, the identity is proved for all components.

- $\mathbf{A} \cdot (\mathbf{B} \times \mathbf{C}) = \mathbf{C} \cdot (\mathbf{A} \times \mathbf{B}) = \mathbf{B} \cdot (\mathbf{C} \times \mathbf{A})$:

$$\mathbf{A} \cdot (\mathbf{B} \times \mathbf{C}) = \epsilon_{ijk} A_i B_j C_k \qquad \text{(Eq. 495)}$$
$$= \epsilon_{kij} A_i B_j C_k \qquad \text{(Eq. 180)}$$
$$= \epsilon_{kij} C_k A_i B_j \qquad \text{(commutativity)}$$
$$= \mathbf{C} \cdot (\mathbf{A} \times \mathbf{B}) \qquad \text{(Eq. 495)}$$
$$= \epsilon_{jki} A_i B_j C_k \qquad \text{(Eq. 180)}$$
$$= \epsilon_{jki} B_j C_k A_i \qquad \text{(commutativity)}$$
$$= \mathbf{B} \cdot (\mathbf{C} \times \mathbf{A}) \qquad \text{(Eq. 495)}$$

The negative permutations of this identity can be similarly obtained and proved by changing the order of the vectors in the cross products which results in a sign change.

- $\mathbf{A} \times (\mathbf{B} \times \mathbf{C}) = \mathbf{B}(\mathbf{A} \cdot \mathbf{C}) - \mathbf{C}(\mathbf{A} \cdot \mathbf{B})$:

$$[\mathbf{A} \times (\mathbf{B} \times \mathbf{C})]_i = \epsilon_{ijk} A_j [\mathbf{B} \times \mathbf{C}]_k \qquad \text{(Eq. 494)}$$
$$= \epsilon_{ijk} A_j \epsilon_{klm} B_l C_m \qquad \text{(Eq. 494)}$$
$$= \epsilon_{ijk} \epsilon_{klm} A_j B_l C_m \qquad \text{(commutativity)}$$
$$= \epsilon_{ijk} \epsilon_{lmk} A_j B_l C_m \qquad \text{(Eq. 180)}$$
$$= (\delta_{il}\delta_{jm} - \delta_{im}\delta_{jl}) A_j B_l C_m \qquad \text{(Eq. 554)}$$

7.1.5 Examples of Using Tensor Techniques to Prove Identities

$$\begin{aligned}
&= \delta_{il}\delta_{jm}A_jB_lC_m - \delta_{im}\delta_{jl}A_jB_lC_m &&\text{(distributivity)}\\
&= (\delta_{il}B_l)(\delta_{jm}A_jC_m) - (\delta_{im}C_m)(\delta_{jl}A_jB_l) &&\text{(commutativity \& grouping)}\\
&= B_i(A_mC_m) - C_i(A_lB_l) &&\text{(Eq. 555)}\\
&= B_i(\mathbf{A}\cdot\mathbf{C}) - C_i(\mathbf{A}\cdot\mathbf{B}) &&\text{(Eq. 493)}\\
&= [\mathbf{B}(\mathbf{A}\cdot\mathbf{C})]_i - [\mathbf{C}(\mathbf{A}\cdot\mathbf{B})]_i &&\text{(definition of index)}\\
&= [\mathbf{B}(\mathbf{A}\cdot\mathbf{C}) - \mathbf{C}(\mathbf{A}\cdot\mathbf{B})]_i &&\text{(Eq. 24)}
\end{aligned}$$

Because i is a free index, the identity is proved for all components. The other variant of this identity, i.e. $(\mathbf{A}\times\mathbf{B})\times\mathbf{C}$, can be obtained and proved similarly by changing the order of the factors in the external cross product with adding a minus sign.

- $\mathbf{A}\times(\nabla\times\mathbf{B}) = (\nabla\mathbf{B})\cdot\mathbf{A} - \mathbf{A}\cdot\nabla\mathbf{B}$:

$$\begin{aligned}
[\mathbf{A}\times(\nabla\times\mathbf{B})]_i &= \epsilon_{ijk}A_j[\nabla\times\mathbf{B}]_k &&\text{(Eq. 494)}\\
&= \epsilon_{ijk}A_j\epsilon_{klm}\partial_lB_m &&\text{(Eq. 423)}\\
&= \epsilon_{ijk}\epsilon_{klm}A_j\partial_lB_m &&\text{(commutativity)}\\
&= \epsilon_{ijk}\epsilon_{lmk}A_j\partial_lB_m &&\text{(Eq. 180)}\\
&= (\delta_{il}\delta_{jm} - \delta_{im}\delta_{jl})A_j\partial_lB_m &&\text{(Eq. 554)}\\
&= \delta_{il}\delta_{jm}A_j\partial_lB_m - \delta_{im}\delta_{jl}A_j\partial_lB_m &&\text{(distributivity)}\\
&= A_m\partial_iB_m - A_l\partial_lB_i &&\text{(Eq. 555)}\\
&= (\partial_iB_m)A_m - A_l(\partial_lB_i) &&\text{(commutativity \& grouping)}\\
&= [(\nabla\mathbf{B})\cdot\mathbf{A}]_i - [\mathbf{A}\cdot\nabla\mathbf{B}]_i &&\text{(Eq. 418 \& § 3.2.5)}\\
&= [(\nabla\mathbf{B})\cdot\mathbf{A} - \mathbf{A}\cdot\nabla\mathbf{B}]_i &&\text{(Eq. 24)}
\end{aligned}$$

Because i is a free index, the identity is proved for all components.

- $\nabla\times(\nabla\times\mathbf{A}) = \nabla(\nabla\cdot\mathbf{A}) - \nabla^2\mathbf{A}$:

$$\begin{aligned}
[\nabla\times(\nabla\times\mathbf{A})]_i &= \epsilon_{ijk}\partial_j[\nabla\times\mathbf{A}]_k &&\text{(Eq. 423)}\\
&= \epsilon_{ijk}\partial_j(\epsilon_{klm}\partial_lA_m) &&\text{(Eq. 423)}\\
&= \epsilon_{ijk}\epsilon_{klm}\partial_j(\partial_lA_m) &&\text{(∂ not acting on ϵ)}\\
&= \epsilon_{ijk}\epsilon_{lmk}\partial_j\partial_lA_m &&\text{(Eq. 180 \& definition of derivative)}\\
&= (\delta_{il}\delta_{jm} - \delta_{im}\delta_{jl})\partial_j\partial_lA_m &&\text{(Eq. 554)}\\
&= \delta_{il}\delta_{jm}\partial_j\partial_lA_m - \delta_{im}\delta_{jl}\partial_j\partial_lA_m &&\text{(distributivity)}\\
&= \partial_m\partial_iA_m - \partial_l\partial_lA_i &&\text{(Eq. 555)}\\
&= \partial_i(\partial_mA_m) - \partial_{ll}A_i &&\text{(∂ shift, grouping \& Eq. 2)}\\
&= [\nabla(\nabla\cdot\mathbf{A})]_i - [\nabla^2\mathbf{A}]_i &&\text{(Eqs. 419, 417 \& 426)}\\
&= [\nabla(\nabla\cdot\mathbf{A}) - \nabla^2\mathbf{A}]_i &&\text{(Eq. 24)}
\end{aligned}$$

Because i is a free index, the identity is proved for all components. This identity can also be considered as an instance of the identity before the last one, observing that in the second term on the right hand side the Laplacian should precede the vector, and hence no independent proof is required.

7.1.5 Examples of Using Tensor Techniques to Prove Identities

- $\nabla (\mathbf{A} \cdot \mathbf{B}) = \mathbf{A} \times (\nabla \times \mathbf{B}) + \mathbf{B} \times (\nabla \times \mathbf{A}) + (\mathbf{A} \cdot \nabla) \mathbf{B} + (\mathbf{B} \cdot \nabla) \mathbf{A}$:

We start from the right hand side and end with the left hand side:

$$[\mathbf{A} \times (\nabla \times \mathbf{B}) + \mathbf{B} \times (\nabla \times \mathbf{A}) + (\mathbf{A} \cdot \nabla) \mathbf{B} + (\mathbf{B} \cdot \nabla) \mathbf{A}]_i =$$

$$[\mathbf{A} \times (\nabla \times \mathbf{B})]_i + [\mathbf{B} \times (\nabla \times \mathbf{A})]_i + [(\mathbf{A} \cdot \nabla) \mathbf{B}]_i + [(\mathbf{B} \cdot \nabla) \mathbf{A}]_i =$$
(Eq. 24)

$$\epsilon_{ijk} A_j [\nabla \times \mathbf{B}]_k + \epsilon_{ijk} B_j [\nabla \times \mathbf{A}]_k + (A_l \partial_l) B_i + (B_l \partial_l) A_i =$$
(Eqs. 494, 419 & indexing)

$$\epsilon_{ijk} A_j (\epsilon_{klm} \partial_l B_m) + \epsilon_{ijk} B_j (\epsilon_{klm} \partial_l A_m) + (A_l \partial_l) B_i + (B_l \partial_l) A_i =$$
(Eq. 423)

$$\epsilon_{ijk} \epsilon_{klm} A_j \partial_l B_m + \epsilon_{ijk} \epsilon_{klm} B_j \partial_l A_m + (A_l \partial_l) B_i + (B_l \partial_l) A_i =$$
(commutativity)

$$\epsilon_{ijk} \epsilon_{lmk} A_j \partial_l B_m + \epsilon_{ijk} \epsilon_{lmk} B_j \partial_l A_m + (A_l \partial_l) B_i + (B_l \partial_l) A_i =$$
(Eq. 180)

$$(\delta_{il}\delta_{jm} - \delta_{im}\delta_{jl}) A_j \partial_l B_m + (\delta_{il}\delta_{jm} - \delta_{im}\delta_{jl}) B_j \partial_l A_m + (A_l \partial_l) B_i + (B_l \partial_l) A_i =$$
(Eq. 554)

$$(\delta_{il}\delta_{jm} A_j \partial_l B_m - \delta_{im}\delta_{jl} A_j \partial_l B_m) + (\delta_{il}\delta_{jm} B_j \partial_l A_m - \delta_{im}\delta_{jl} B_j \partial_l A_m)$$
$$+ (A_l \partial_l) B_i + (B_l \partial_l) A_i =$$
(distributivity)

$$\delta_{il}\delta_{jm} A_j \partial_l B_m - A_l \partial_l B_i + \delta_{il}\delta_{jm} B_j \partial_l A_m - B_l \partial_l A_i + (A_l \partial_l) B_i + (B_l \partial_l) A_i =$$
(Eq. 555)

$$\delta_{il}\delta_{jm} A_j \partial_l B_m - (A_l \partial_l) B_i + \delta_{il}\delta_{jm} B_j \partial_l A_m - (B_l \partial_l) A_i + (A_l \partial_l) B_i + (B_l \partial_l) A_i =$$
(grouping)

$$\delta_{il}\delta_{jm} A_j \partial_l B_m + \delta_{il}\delta_{jm} B_j \partial_l A_m =$$
(cancellation)

$$A_m \partial_i B_m + B_m \partial_i A_m =$$
(Eq. 555)

$$\partial_i (A_m B_m) =$$
(product rule)

$$[\nabla (\mathbf{A} \cdot \mathbf{B})]_i$$
(Eqs. 417 & 419)

Because i is a free index, the identity is proved for all components.

- $\nabla \cdot (\mathbf{A} \times \mathbf{B}) = \mathbf{B} \cdot (\nabla \times \mathbf{A}) - \mathbf{A} \cdot (\nabla \times \mathbf{B})$:

$$\nabla \cdot (\mathbf{A} \times \mathbf{B}) = \partial_i [\mathbf{A} \times \mathbf{B}]_i \qquad \text{(Eq. 419)}$$
$$= \partial_i (\epsilon_{ijk} A_j B_k) \qquad \text{(Eq. 494)}$$
$$= \epsilon_{ijk} \partial_i (A_j B_k) \qquad (\partial \text{ not acting on } \epsilon)$$
$$= \epsilon_{ijk} (B_k \partial_i A_j + A_j \partial_i B_k) \qquad \text{(product rule)}$$

7.1.5 Examples of Using Tensor Techniques to Prove Identities

$$\begin{aligned}
&= \epsilon_{ijk} B_k \partial_i A_j + \epsilon_{ijk} A_j \partial_i B_k && \text{(distributivity)} \\
&= \epsilon_{kij} B_k \partial_i A_j - \epsilon_{jik} A_j \partial_i B_k && \text{(Eq. 180)} \\
&= B_k \left(\epsilon_{kij} \partial_i A_j \right) - A_j \left(\epsilon_{jik} \partial_i B_k \right) && \text{(commutativity \& grouping)} \\
&= B_k \left[\nabla \times \mathbf{A} \right]_k - A_j \left[\nabla \times \mathbf{B} \right]_j && \text{(Eq. 423)} \\
&= \mathbf{B} \cdot (\nabla \times \mathbf{A}) - \mathbf{A} \cdot (\nabla \times \mathbf{B}) && \text{(Eq. 493)}
\end{aligned}$$

- $\nabla \times (\mathbf{A} \times \mathbf{B}) = (\mathbf{B} \cdot \nabla) \mathbf{A} + (\nabla \cdot \mathbf{B}) \mathbf{A} - (\nabla \cdot \mathbf{A}) \mathbf{B} - (\mathbf{A} \cdot \nabla) \mathbf{B}$:

$$\begin{aligned}
\left[\nabla \times (\mathbf{A} \times \mathbf{B}) \right]_i &= \epsilon_{ijk} \partial_j \left[\mathbf{A} \times \mathbf{B} \right]_k && \text{(Eq. 423)} \\
&= \epsilon_{ijk} \partial_j \left(\epsilon_{klm} A_l B_m \right) && \text{(Eq. 494)} \\
&= \epsilon_{ijk} \epsilon_{klm} \partial_j \left(A_l B_m \right) && (\partial \text{ not acting on } \epsilon) \\
&= \epsilon_{ijk} \epsilon_{klm} \left(B_m \partial_j A_l + A_l \partial_j B_m \right) && \text{(product rule)} \\
&= \epsilon_{ijk} \epsilon_{lmk} \left(B_m \partial_j A_l + A_l \partial_j B_m \right) && \text{(Eq. 180)} \\
&= \left(\delta_{il} \delta_{jm} - \delta_{im} \delta_{jl} \right) \left(B_m \partial_j A_l + A_l \partial_j B_m \right) && \text{(Eq. 554)} \\
&= \delta_{il} \delta_{jm} B_m \partial_j A_l + \delta_{il} \delta_{jm} A_l \partial_j B_m - \\
&\quad \delta_{im} \delta_{jl} B_m \partial_j A_l - \delta_{im} \delta_{jl} A_l \partial_j B_m && \text{(distributivity)} \\
&= B_m \partial_m A_i + A_i \partial_m B_m - B_i \partial_j A_j - A_j \partial_j B_i && \text{(Eq. 555)} \\
&= \left(B_m \partial_m \right) A_i + \left(\partial_m B_m \right) A_i - \left(\partial_j A_j \right) B_i - \left(A_j \partial_j \right) B_i && \text{(grouping)} \\
&= \left[(\mathbf{B} \cdot \nabla) \mathbf{A} \right]_i + \left[(\nabla \cdot \mathbf{B}) \mathbf{A} \right]_i - \\
&\quad \left[(\nabla \cdot \mathbf{A}) \mathbf{B} \right]_i - \left[(\mathbf{A} \cdot \nabla) \mathbf{B} \right]_i && \text{(Eqs. 413 \& 419)} \\
&= \left[(\mathbf{B} \cdot \nabla) \mathbf{A} + (\nabla \cdot \mathbf{B}) \mathbf{A} - (\nabla \cdot \mathbf{A}) \mathbf{B} - (\mathbf{A} \cdot \nabla) \mathbf{B} \right]_i && \text{(Eq. 24)}
\end{aligned}$$

Because i is a free index, the identity is proved for all components.

- $(\mathbf{A} \times \mathbf{B}) \cdot (\mathbf{C} \times \mathbf{D}) = \begin{vmatrix} \mathbf{A} \cdot \mathbf{C} & \mathbf{A} \cdot \mathbf{D} \\ \mathbf{B} \cdot \mathbf{C} & \mathbf{B} \cdot \mathbf{D} \end{vmatrix}$:

$$\begin{aligned}
(\mathbf{A} \times \mathbf{B}) \cdot (\mathbf{C} \times \mathbf{D}) &= \left[\mathbf{A} \times \mathbf{B} \right]_i \left[\mathbf{C} \times \mathbf{D} \right]_i && \text{(Eq. 493)} \\
&= \epsilon_{ijk} A_j B_k \epsilon_{ilm} C_l D_m && \text{(Eq. 494)} \\
&= \epsilon_{ijk} \epsilon_{ilm} A_j B_k C_l D_m && \text{(commutativity)} \\
&= \left(\delta_{jl} \delta_{km} - \delta_{jm} \delta_{kl} \right) A_j B_k C_l D_m && \text{(Eqs. 180 \& 554)} \\
&= \delta_{jl} \delta_{km} A_j B_k C_l D_m - \delta_{jm} \delta_{kl} A_j B_k C_l D_m && \text{(distributivity)} \\
&= \left(\delta_{jl} A_j C_l \right) \left(\delta_{km} B_k D_m \right) - \left(\delta_{jm} A_j D_m \right) \left(\delta_{kl} B_k C_l \right) && \text{(commutativity)} \\
&= \left(A_l C_l \right) \left(B_m D_m \right) - \left(A_m D_m \right) \left(B_l C_l \right) && \text{(Eq. 555)} \\
&= (\mathbf{A} \cdot \mathbf{C}) (\mathbf{B} \cdot \mathbf{D}) - (\mathbf{A} \cdot \mathbf{D}) (\mathbf{B} \cdot \mathbf{C}) && \text{(Eq. 493)} \\
&= \begin{vmatrix} \mathbf{A} \cdot \mathbf{C} & \mathbf{A} \cdot \mathbf{D} \\ \mathbf{B} \cdot \mathbf{C} & \mathbf{B} \cdot \mathbf{D} \end{vmatrix} && \text{(definition)}
\end{aligned}$$

- $(\mathbf{A} \times \mathbf{B}) \times (\mathbf{C} \times \mathbf{D}) = \left[\mathbf{D} \cdot (\mathbf{A} \times \mathbf{B}) \right] \mathbf{C} - \left[\mathbf{C} \cdot (\mathbf{A} \times \mathbf{B}) \right] \mathbf{D}$:

$$\begin{aligned}
\left[(\mathbf{A} \times \mathbf{B}) \times (\mathbf{C} \times \mathbf{D}) \right]_i &= \epsilon_{ijk} \left[\mathbf{A} \times \mathbf{B} \right]_j \left[\mathbf{C} \times \mathbf{D} \right]_k && \text{(Eq. 494)} \\
&= \epsilon_{ijk} \epsilon_{jmn} A_m B_n \epsilon_{kpq} C_p D_q && \text{(Eq. 494)}
\end{aligned}$$

$$
\begin{aligned}
&= \epsilon_{ijk}\epsilon_{kpq}\epsilon_{jmn}A_m B_n C_p D_q && \text{(commutativity)} \\
&= \epsilon_{ijk}\epsilon_{pqk}\epsilon_{jmn}A_m B_n C_p D_q && \text{(Eq. 180)} \\
&= (\delta_{ip}\delta_{jq} - \delta_{iq}\delta_{jp})\epsilon_{jmn}A_m B_n C_p D_q && \text{(Eq. 554)} \\
&= (\delta_{ip}\delta_{jq}\epsilon_{jmn} - \delta_{iq}\delta_{jp}\epsilon_{jmn})A_m B_n C_p D_q && \text{(distributivity)} \\
&= (\delta_{ip}\epsilon_{qmn} - \delta_{iq}\epsilon_{pmn})A_m B_n C_p D_q && \text{(Eq. 555)} \\
&= \delta_{ip}\epsilon_{qmn}A_m B_n C_p D_q - \delta_{iq}\epsilon_{pmn}A_m B_n C_p D_q && \text{(distributivity)} \\
&= \epsilon_{qmn}A_m B_n C_i D_q - \epsilon_{pmn}A_m B_n C_p D_i && \text{(Eq. 555)} \\
&= \epsilon_{qmn}D_q A_m B_n C_i - \epsilon_{pmn}C_p A_m B_n D_i && \text{(commutativity)} \\
&= (\epsilon_{qmn}D_q A_m B_n)C_i - (\epsilon_{pmn}C_p A_m B_n)D_i && \text{(grouping)} \\
&= [\mathbf{D}\cdot(\mathbf{A}\times\mathbf{B})]C_i - [\mathbf{C}\cdot(\mathbf{A}\times\mathbf{B})]D_i && \text{(Eq. 495)} \\
&= [[\mathbf{D}\cdot(\mathbf{A}\times\mathbf{B})]\mathbf{C}]_i - [[\mathbf{C}\cdot(\mathbf{A}\times\mathbf{B})]\mathbf{D}]_i && \text{(index definition)} \\
&= [[\mathbf{D}\cdot(\mathbf{A}\times\mathbf{B})]\mathbf{C} - [\mathbf{C}\cdot(\mathbf{A}\times\mathbf{B})]\mathbf{D}]_i && \text{(Eq. 24)}
\end{aligned}
$$

Because i is a free index, the identity is proved for all components.

7.2 Tensors in Geometry

In this section, we give some examples of tensor calculus applications in geometry. These examples are related mainly to the investigation of the properties of spaces in general and space curvature in particular and hence they play important roles in non-Euclidean geometries and related applications in the physical sciences such as geometry-based gravitational theories.

7.2.1 Riemann-Christoffel Curvature Tensor

This absolute rank-4 tensor, which is also called Riemann curvature tensor and Riemann-Christoffel tensor, is a property of the space. It characterizes important properties of spaces and hence it plays an important role in geometry in general and in non-Euclidean geometries in particular with many applications in geometry-based physical theories. The tensor is used, for instance, to test for the space flatness (see § 2.1). As indicated before, the Riemann-Christoffel curvature tensor vanishes identically *iff* the space is globally flat and hence the Riemann-Christoffel curvature tensor is zero in Euclidean spaces. The Riemann-Christoffel curvature tensor depends only on the metric which, in general coordinate systems, is a function of position and hence the Riemann-Christoffel curvature tensor follows this dependency on position. Yes, for affine coordinate systems the metric tensor is constant and hence the Riemann-Christoffel curvature tensor vanishes identically.

There are two kinds of Riemann-Christoffel curvature tensor: first and second, where the first kind is a type $(0,4)$ tensor while the second kind is a type $(1,3)$ tensor. The Riemann-Christoffel curvature tensor of the first kind, which may also be called the covariant (or totally covariant) Riemann-Christoffel curvature tensor, is defined by:

$$R_{ijkl} = \partial_k[jl,i] - \partial_l[jk,i] + [il,r]\Gamma^r_{jk} - [ik,r]\Gamma^r_{jl} \tag{558}$$

7.2.1 Riemann-Christoffel Curvature Tensor

$$= \frac{1}{2} \left(\partial_k \partial_j g_{li} + \partial_l \partial_i g_{jk} - \partial_k \partial_i g_{jl} - \partial_l \partial_j g_{ki} \right) + [il, r] \Gamma^r_{jk} - [ik, r] \Gamma^r_{jl}$$

$$= \frac{1}{2} \left(\partial_k \partial_j g_{li} + \partial_l \partial_i g_{jk} - \partial_k \partial_i g_{jl} - \partial_l \partial_j g_{ki} \right) + g^{rs} \left([il, r][jk, s] - [ik, r][jl, s] \right)$$

where the second step is based on Eq. 355 while the third step is based on Eq. 308. We note that the first line of Eq. 558 can be cast in the following mnemonic determinantal form:

$$R_{ijkl} = \begin{vmatrix} \partial_k & \partial_l \\ [jk, i] & [jl, i] \end{vmatrix} + \begin{vmatrix} \Gamma^r_{jk} & \Gamma^r_{jl} \\ [ik, r] & [il, r] \end{vmatrix} \tag{559}$$

Similarly, the Riemann-Christoffel curvature tensor of the second kind, which may also be called the mixed Riemann-Christoffel curvature tensor, is given by:

$$R^i_{jkl} = \partial_k \Gamma^i_{jl} - \partial_l \Gamma^i_{jk} + \Gamma^r_{jl} \Gamma^i_{rk} - \Gamma^r_{jk} \Gamma^i_{rl} \tag{560}$$

We note again that Eq. 560 can be put into the following mnemonic determinantal form:

$$R^i{}_{jkl} = \begin{vmatrix} \partial_k & \partial_l \\ \Gamma^i_{jk} & \Gamma^i_{jl} \end{vmatrix} + \begin{vmatrix} \Gamma^r_{jl} & \Gamma^r_{jk} \\ \Gamma^i_{rl} & \Gamma^i_{rk} \end{vmatrix} \tag{561}$$

On lowering the contravariant index of the Riemann-Christoffel curvature tensor of the second kind, the Riemann-Christoffel curvature tensor of the first kind is obtained, that is:

$$R_{ijkl} = g_{ia} R^a{}_{jkl} \tag{562}$$

Similarly, the Riemann-Christoffel curvature tensor of the second kind can be obtained by raising the first index of the Riemann-Christoffel curvature tensor of the first kind, that is:

$$R^i{}_{jkl} = g^{ia} R_{ajkl} \tag{563}$$

One of the main applications of the Riemann-Christoffel curvature tensor in tensor calculus is demonstrated in its role in tensor differentiation. As seen in § 5.2, the covariant differential operators in mixed derivatives are not commutative and hence for a covariant vector **A** we have:

$$A_{j;kl} - A_{j;lk} = R^i{}_{jkl} A_i \tag{564}$$

From Eq. 564, it is obvious that the mixed second order covariant derivatives are equal *iff* the Riemann-Christoffel curvature tensor of the second kind vanishes identically which means that the space is flat and hence the covariant derivatives are ordinary partial derivatives. Similarly, for the mixed second order covariant derivative of a contravariant vector **A** we have:

$$A^j{}_{;kl} - A^j{}_{;lk} = R^j{}_{ilk} A^i \tag{565}$$

which is similar to Eq. 564 for a covariant vector **A**. In brief, the covariant differential operators become commutative when the metric makes the Riemann-Christoffel curvature tensor of either kind vanish. So, since the Riemann-Christoffel curvature tensor is zero in

7.2.1 Riemann-Christoffel Curvature Tensor

Euclidean spaces, the mixed second order covariant derivatives, which become ordinary partial derivatives, are equal when the C^2 continuity condition is satisfied.

The Riemann-Christoffel curvature tensor of the first kind satisfies the following symmetric and skew-symmetric relations in its four indices:

$$R_{ijkl} = R_{klij} \quad \text{(block symmetry)} \tag{566}$$
$$= -R_{jikl} \quad \text{(anti-symmetry in the first two indices)}$$
$$= -R_{ijlk} \quad \text{(anti-symmetry in the last two indices)}$$

The skew-symmetric property of the covariant Riemann-Christoffel curvature tensor with respect to the last two indices also applies to the mixed Riemann-Christoffel curvature tensor, that is:

$$R^i{}_{jkl} = -R^i{}_{jlk} \tag{567}$$

As a consequence of the two anti-symmetric properties of the covariant Riemann-Christoffel curvature tensor, the entries of the Riemann-Christoffel curvature tensor with identical values of the first two indices (e.g. R_{iijk}) or/and the last two indices (e.g. R_{ijkk}) are zero (refer to § 3.1.5). Also, as a consequence of these anti-symmetric properties, all the entries of the tensor with identical values of more than two indices (e.g. R_{iiji}) are zero.

We remark that all the above symmetric and anti-symmetric properties of the Riemann-Christoffel curvature tensor of the first and second kinds can be proved by using the above definitions of this tensor. For example, the first anti-symmetric property can be verified as follows:

$$R_{jikl} = \frac{1}{2} \left(\partial_k \partial_i g_{lj} + \partial_l \partial_j g_{ik} - \partial_k \partial_j g_{il} - \partial_l \partial_i g_{kj} \right) + g^{rs} \left([jl, r][ik, s] - [jk, r][il, s] \right) \tag{568}$$
$$= -\left[\frac{1}{2} \left(\partial_k \partial_j g_{il} + \partial_l \partial_i g_{kj} - \partial_k \partial_i g_{lj} - \partial_l \partial_j g_{ik} \right) + g^{rs} \left([jk, r][il, s] - [jl, r][ik, s] \right) \right]$$
$$= -\left[\frac{1}{2} \left(\partial_k \partial_j g_{li} + \partial_l \partial_i g_{jk} - \partial_k \partial_i g_{jl} - \partial_l \partial_j g_{ki} \right) + g^{sr} \left([il, s][jk, r] - [ik, s][jl, r] \right) \right]$$
$$= -\left[\frac{1}{2} \left(\partial_k \partial_j g_{li} + \partial_l \partial_i g_{jk} - \partial_k \partial_i g_{jl} - \partial_l \partial_j g_{ki} \right) + g^{rs} \left([il, r][jk, s] - [ik, r][jl, s] \right) \right]$$
$$= -R_{ijkl}$$

where in the first line we use the third line of Eq. 558 with exchanging i and j, in the second line we take a factor of -1, in the third line we use the symmetry of the metric tensor in its two indices, in the fourth line we relabel two dummy indices, and in the last line we use the third line of Eq. 558 again.

In an nD space, the Riemann-Christoffel curvature tensor has n^4 components. As a consequence of the symmetric and anti-symmetric properties of the covariant Riemann-Christoffel curvature tensor, there are three types of distinct non-vanishing entries:

1. Entries with only two distinct indices (type R_{ijij}) which count:

$$N_2 = \frac{n(n-1)}{2} \tag{569}$$

7.2.1 Riemann-Christoffel Curvature Tensor

2. Entries with only three distinct indices (type R_{ijik}) which count:

$$N_3 = \frac{n(n-1)(n-2)}{2} \tag{570}$$

3. Entries with four distinct indices (type R_{ijkl}) which count:

$$N_4 = \frac{n(n-1)(n-2)(n-3)}{12} \tag{571}$$

The numerator in these three equations represents the k-permutations of n distinct objects which is given by:

$$P(n,k) = \frac{n!}{(n-k)!} \tag{572}$$

where k in these cases is equal to $2, 3, 4$ respectively, while the denominator represents the number of non-independent ways of generating these entries due to the anti-symmetric and block symmetric properties.

By adding the numbers of the three types of non-zero distinct entries, as given by Eqs. 569-571, we can see that the Riemann-Christoffel curvature tensor in an nD space has a total of:

$$N_{\text{RI}} = N_2 + N_3 + N_4 = \frac{n^2(n^2-1)}{12} \tag{573}$$

independent components which do not vanish identically. For example, in a 2D Riemannian space the Riemann-Christoffel curvature tensor has $2^4 = 16$ components; however there is only one independent component (with the principal suffix 1212) which is not identically zero represented by the following four interdependent components:

$$R_{1212} = R_{2121} = -R_{1221} = -R_{2112} \tag{574}$$

Similarly, in a 3D Riemannian space the Riemann-Christoffel curvature tensor has $3^4 = 81$ components but only six of these are distinct non-zero entries which are the ones with the following principal suffixes:

$$1212, 1313, 1213, 2123, 3132, 2323 \tag{575}$$

where the permutations of the indices in each one of these suffixes are subject to the symmetric and anti-symmetric properties of the four indices of the Riemann-Christoffel curvature tensor, as in the case of a 2D space in the above example, and hence these permutations do not produce independent entries. Following the pattern in the examples of 2D and 3D spaces, in a 4D Riemannian space the Riemann-Christoffel curvature tensor has $4^4 = 256$ components but only 20 of these are independent non-zero entries, while in a 5D Riemannian space the Riemann-Christoffel curvature tensor has $5^4 = 625$ components but only 50 are independent non-zero entries.

On contracting the first covariant index with the contravariant index of the Riemann-Christoffel curvature tensor of the second kind we obtain:

$$R^i_{ikl} = \partial_k \Gamma^i_{il} - \partial_l \Gamma^i_{ik} + \Gamma^r_{il}\Gamma^i_{rk} - \Gamma^r_{ik}\Gamma^i_{rl} \qquad (j = i \text{ in Eq. 560}) \tag{576}$$

$$\begin{aligned}
&= \partial_k \Gamma^i_{il} - \partial_l \Gamma^i_{ik} + \Gamma^r_{il}\Gamma^i_{rk} - \Gamma^i_{rk}\Gamma^r_{il} &&\text{(relabeling dummy } i,r \text{ in last term)}\\
&= \partial_k \Gamma^i_{il} - \partial_l \Gamma^i_{ik}\\
&= \partial_k \left[\partial_l \left(\ln \sqrt{g}\right)\right] - \partial_l \left[\partial_k \left(\ln \sqrt{g}\right)\right] &&\text{(Eq. 324)}\\
&= \partial_k \partial_l \left(\ln \sqrt{g}\right) - \partial_l \partial_k \left(\ln \sqrt{g}\right)\\
&= \partial_k \partial_l \left(\ln \sqrt{g}\right) - \partial_k \partial_l \left(\ln \sqrt{g}\right) &&(C^2 \text{ continuity condition})\\
&= 0
\end{aligned}$$

That is:
$$R^i_{ikl} = 0 \tag{577}$$

This is inline with the anti-symmetric property of the first two indices of the totally covariant Riemann-Christoffel curvature tensor.

Finally, the Riemann-Christoffel curvature tensor also satisfies the following identity:
$$R_{ijkl;s} + R_{iljk;s} = R_{iksl;j} + R_{ikjs;l} \tag{578}$$

As indicated earlier, a necessary and sufficient condition for a space to have a coordinate system such that all the components of the metric tensor are constants is that:
$$R_{ijkl} = 0 \tag{579}$$

This can be concluded from Eq. 558 plus Eqs. 307 and 308.

7.2.2 Bianchi Identities

The Riemann-Christoffel curvature tensor of the first and second kind satisfies a number of identities called the Bianchi identities. The first Bianchi identity is given by:

$$R_{ijkl} + R_{iljk} + R_{iklj} = 0 \qquad \text{(first kind)} \tag{580}$$
$$R^i{}_{jkl} + R^i{}_{ljk} + R^i{}_{klj} = 0 \qquad \text{(second kind)} \tag{581}$$

These two forms of the first Bianchi identity can be obtained from each other by the index raising and lowering operators. The first Bianchi identity, as stated above, is an instance of the fact that by fixing the position of one of the four indices and permuting the other three indices cyclically, the algebraic sum of these three permuting forms is zero, that is:

$$R_{ijkl} + R_{iljk} + R_{iklj} = 0 \qquad (i \text{ fixed}) \tag{582}$$
$$R_{ijkl} + R_{ljik} + R_{kjli} = 0 \qquad (j \text{ fixed}) \tag{583}$$
$$R_{ijkl} + R_{likj} + R_{jlki} = 0 \qquad (k \text{ fixed}) \tag{584}$$
$$R_{ijkl} + R_{kijl} + R_{jkil} = 0 \qquad (l \text{ fixed}) \tag{585}$$

All the above identities can be easily proved by using the definition of the Riemann-Christoffel curvature tensor noting that the Christoffel symbols of both kinds are symmetric in their paired indices as given by Eqs. 310 and 311. For example, Eq. 580 can be verified by substituting from the first line of Eq. 558 into Eq. 580, that is:

$$R_{ijkl} + R_{iljk} + R_{iklj} \;=\; \partial_k \left[jl,i\right] - \partial_l \left[jk,i\right] + \left[il,r\right]\Gamma^r_{jk} - \left[ik,r\right]\Gamma^r_{jl} + \tag{586}$$

7.2.3 Ricci Curvature Tensor and Scalar 172

$$\partial_j [lk, i] - \partial_k [lj, i] + [ik, r] \Gamma^r_{lj} - [ij, r] \Gamma^r_{lk} +$$
$$\partial_l [kj, i] - \partial_j [kl, i] + [ij, r] \Gamma^r_{kl} - [il, r] \Gamma^r_{kj}$$
$$= 0$$

Another one of the Bianchi identities is:

$$R_{ijkl;m} + R_{ijlm;k} + R_{ijmk;l} = 0 \qquad \text{(first kind)} \qquad (587)$$
$$R^i{}_{jkl;m} + R^i{}_{jlm;k} + R^i{}_{jmk;l} = 0 \qquad \text{(second kind)} \qquad (588)$$

Again, these two forms can be obtained from each other by the raising and lowering operators. We note that the pattern of the second Bianchi identity in its both kinds is simple, that is the first two indices are fixed while the last three indices are cyclically permuted in the three terms. It is noteworthy that the Bianchi identities are valid regardless of the space metric.

7.2.3 Ricci Curvature Tensor and Scalar

The Ricci curvature tensor, which is a rank-2 absolute symmetric tensor, is a byproduct of the Riemann-Christoffel curvature tensor and hence it plays a similar role in characterizing the space and describing its curvature. There are two kinds of Ricci curvature tensor: first and second, where the first kind is a type $(0, 2)$ tensor while the second kind is a type $(1, 1)$ tensor. The first kind of this tensor is obtained by contracting the contravariant index with the last covariant index of the Riemann-Christoffel curvature tensor of the second kind, that is:

$$R_{ij} = R^a{}_{ija} = \partial_j \Gamma^a_{ia} - \partial_a \Gamma^a_{ij} + \Gamma^a_{bj} \Gamma^b_{ia} - \Gamma^a_{ba} \Gamma^b_{ij} \qquad (589)$$

where Eq. 560 is used in this formulation. The Ricci curvature tensor, as given by the last equation, can be written in the following mnemonic determinantal form:

$$R_{ij} = \begin{vmatrix} \partial_j & \partial_a \\ \Gamma^a_{ij} & \Gamma^a_{ia} \end{vmatrix} + \begin{vmatrix} \Gamma^a_{bj} & \Gamma^a_{ba} \\ \Gamma^b_{ij} & \Gamma^b_{ia} \end{vmatrix} \qquad (590)$$

Because of Eq. 324 (i.e. $\Gamma^j_{ij} = \partial_i (\ln \sqrt{g})$), the Ricci tensor can also be written in the following forms as well as several other forms:

$$\begin{aligned} R_{ij} &= \partial_j \partial_i (\ln \sqrt{g}) - \partial_a \Gamma^a_{ij} + \Gamma^a_{bj} \Gamma^b_{ia} - \Gamma^b_{ij} \partial_b (\ln \sqrt{g}) & (591) \\ &= \partial_j \partial_i (\ln \sqrt{g}) + \Gamma^a_{bj} \Gamma^b_{ia} - \partial_a \Gamma^a_{ij} - \Gamma^a_{ij} \partial_a (\ln \sqrt{g}) \\ &= \partial_j \partial_i (\ln \sqrt{g}) + \Gamma^a_{bj} \Gamma^b_{ia} - \frac{1}{\sqrt{g}} \partial_a \left(\sqrt{g} \Gamma^a_{ij} \right) \end{aligned}$$

where g is the determinant of the covariant metric tensor. As stated above, the Ricci tensor of the first kind is symmetric, that is:

$$R_{ij} = R_{ji} \qquad (592)$$

This can be easily verified by exchanging the i and j indices in the last line of Eq. 591 taking account of the C^2 continuity condition and the fact that the Christoffel symbols are symmetric in their paired indices (Eq. 311).

On raising the first index of the Ricci tensor of the first kind, the Ricci tensor of the second kind is obtained, that is:

$$R^i{}_j = g^{ik} R_{kj} \tag{593}$$

This process can be reversed and hence the Ricci tensor of the first kind can be obtained from the Ricci tensor of the second kind using the index lowering operator. In an nD space, the Ricci tensor has n^2 entries. However, because of its symmetry the number of its distinct entries is reduced to:

$$N_{\text{RD}} = \frac{n(n+1)}{2} \tag{594}$$

The Ricci scalar $\mathcal{R}$, which is also called the curvature scalar and the curvature invariant, is the result of contracting the indices of the Ricci tensor of the second kind, that is:

$$\mathcal{R} = R^i{}_i \tag{595}$$

Since the Ricci scalar is obtained by raising a covariant index of the Ricci tensor of the first kind using the raising operator followed by contracting the two indices, it may be written by some in the following form:

$$\mathcal{R} = g^{ij} R_{ij} = g^{ij} \left[\partial_j \partial_i (\ln \sqrt{g}) + \Gamma^a_{bj} \Gamma^b_{ia} - \frac{1}{\sqrt{g}} \partial_a \left(\sqrt{g} \Gamma^a_{ij} \right) \right] \tag{596}$$

where the expression inside the square brackets is obtained from the last line of Eq. 591. Similar expressions can be obtained from the other lines of that equation.

7.3 Tensors in Science

In this section, we give a few examples of tensor calculus applications in science. These examples come mainly from the disciplines of continuum mechanics and fluid dynamics. For simplicity, clarity and widespread use we employ a Cartesian approach in the following formulations.

7.3.1 Infinitesimal Strain Tensor

This is a rank-2 tensor which describes the state of strain in a continuum medium and hence it is used in continuum and fluid mechanics. The infinitesimal strain tensor $\boldsymbol{\gamma}$ is defined by:

$$\boldsymbol{\gamma} = \frac{\nabla \mathbf{d} + (\nabla \mathbf{d})^T}{2} \tag{597}$$

where $\mathbf{d}$ is the displacement vector and the superscript T represents matrix transposition. We note that some authors do not include the factor $\frac{1}{2}$ in the definition of $\boldsymbol{\gamma}$. The displacement vector $\mathbf{d}$ represents the change in distance and direction which an infinitesimal element of the medium experiences as a consequence of the applied stress. In Cartesian coordinates, the last equation is given in tensor notation by the following form:

$$\gamma_{ij} = \frac{\partial_i d_j + \partial_j d_i}{2} \tag{598}$$

7.3.2 Stress Tensor

The stress tensor, which is also called Cauchy stress tensor, is a rank-2 tensor used for *transforming* a normal vector to a surface *to* a traction vector acting on that surface, that is:

$$\mathbf{T} = \boldsymbol{\sigma} \cdot \mathbf{n} \tag{599}$$

where $\mathbf{T}$ is the traction vector, $\boldsymbol{\sigma}$ is the stress tensor and $\mathbf{n}$ is the normal vector. This is usually expressed in tensor notation using Cartesian coordinates by the following form:

$$T_i = \sigma_{ij} n_j \tag{600}$$

We should remark that the stress tensor is symmetric in many applications (e.g. in the flow of Newtonian fluids) but not all, as it can be asymmetric in some cases. We also remark that we chose to define the stress tensor within the context of Cauchy stress law which is more relevant to the continuum mechanics, although it can be defined differently in other disciplines and in a more general form.

The diagonal components of the stress tensor represent normal stresses while the off-diagonal components represent shear stresses. Assuming that the stress tensor is symmetric, it possesses $\frac{n(n+1)}{2}$ independent components, instead of n^2, where n is the space dimension. Hence in a 3D space (which is the natural space for this tensor in the common physical applications) it has six independent components. In fluid dynamics, the stress tensor (which may also be labeled as the total stress tensor) is decomposed into two main parts: a viscous part and a pressure part. The viscous part may then be split into a normal stress and a shear stress while the pressure part may be split into a hydrostatic pressure and an extra pressure of hydrodynamic nature.

7.3.3 Displacement Gradient Tensors

These are rank-2 tensors denoted by $\mathbf{E}$ and $\boldsymbol{\Delta}$. They are defined in Cartesian coordinates using tensor notation as:

$$E_{ij} = \frac{\partial x_i}{\partial x'_j} \qquad \Delta_{ij} = \frac{\partial x'_i}{\partial x_j} \tag{601}$$

where the indexed x and x' represent the Cartesian coordinates of an observed continuum particle at the present and past times respectively. These tensors may also be called the deformation gradient tensors. The first displacement gradient tensor $\mathbf{E}$ quantifies the displacement of a particle at the present time relative to its position at the past time, while the second displacement gradient tensor $\boldsymbol{\Delta}$ quantifies its displacement at the past

7.3.4 Finger Strain Tensor

time relative to its position at the present time. From their definitions, it is obvious that $\mathbf{E}$ and $\mathbf{\Delta}$ are inverses of each other and hence:

$$E_{ik}\Delta_{kj} = \delta_{ij} \tag{602}$$

7.3.4 Finger Strain Tensor

This rank-2 tensor, which may also be called the left Cauchy-Green deformation tensor, is used in the fields of fluid and continuum mechanics to describe the strain in a continuum object, e.g. fluid, in a series of time frames. It is defined as:

$$\mathbf{B} = \mathbf{E} \cdot \mathbf{E}^T \tag{603}$$

where $\mathbf{E}$ is the first displacement gradient tensor as defined in § 7.3.3 and the superscript T represents matrix transposition. The last equation can be expressed in tensor notation, using Cartesian coordinates, as follows:

$$B_{ij} = \frac{\partial x_i}{\partial x'_k}\frac{\partial x_j}{\partial x'_k} \tag{604}$$

where the indexed x and x' represent the Cartesian coordinates of an element of the continuum at the present and past times respectively.

7.3.5 Cauchy Strain Tensor

This tensor, which may also be called the right Cauchy-Green deformation tensor, is the inverse of the Finger strain tensor (see § 7.3.4) and hence it is denoted by $\mathbf{B}^{-1}$. Consequently, it is defined as:

$$\mathbf{B}^{-1} = \mathbf{\Delta}^T \cdot \mathbf{\Delta} \tag{605}$$

where $\mathbf{\Delta}$ is the second displacement gradient tensor as defined in § 7.3.3. The last equation can be expressed in tensor notation, using Cartesian coordinates, as follows:

$$B^{-1}_{ij} = \frac{\partial x'_k}{\partial x_i}\frac{\partial x'_k}{\partial x_j} \tag{606}$$

The Finger and Cauchy strain tensors may be called "finite strain tensors" as opposite to infinitesimal strain tensor (see § 7.3.1). They are symmetric positive definite tensors; moreover they become the unity tensor when the change in the state of the object from the past to the present times consists of rotation and translation with no deformation.

7.3.6 Velocity Gradient Tensor

This is a rank-2 tensor which is often used in fluid dynamics and rheology. As its name suggests, it is the gradient of the velocity vector $\mathbf{v}$ and hence it is given in Cartesian coordinates by:

$$[\nabla \mathbf{v}]_{ij} = \partial_i v_j \tag{607}$$

The velocity gradient tensor in other coordinate systems can be obtained from the expression of the gradient of vectors in these systems, as given, for instance, in § 6.4 and § 6.5 for cylindrical and spherical coordinates. The term "velocity gradient tensor" may also be used for the transpose of this tensor, i.e. $(\nabla \mathbf{v})^T$. The velocity gradient tensor is usually decomposed into a symmetric part which is the rate of strain tensor $\mathbf{S}$ (see § 7.3.7), and an anti-symmetric part which is the vorticity tensor $\bar{\mathbf{S}}$ (see § 7.3.8), that is:

$$\nabla \mathbf{v} = \mathbf{S} + \bar{\mathbf{S}} \tag{608}$$

As seen earlier (refer to § 3.1.5), any rank-2 tensor can be decomposed into a symmetric part and an anti-symmetric part.

7.3.7 Rate of Strain Tensor

This rank-2 tensor, which is also called the rate of deformation tensor, is the symmetric part of the velocity gradient tensor and hence it is given by:

$$\mathbf{S} = \frac{\nabla \mathbf{v} + (\nabla \mathbf{v})^T}{2} \tag{609}$$

which, in tensor notation with Cartesian coordinates, becomes (see Eq. 99):

$$S_{ij} = \frac{\partial_i v_j + \partial_j v_i}{2} \tag{610}$$

We note that some authors do not include the factor $\frac{1}{2}$ in the definition of $\mathbf{S}$ and $\bar{\mathbf{S}}$ and hence this factor is moved to the definition of $\nabla \mathbf{v}$ (Eq. 608). Also, the tensors $\mathbf{S}$ and $\bar{\mathbf{S}}$ are commonly denoted by $\dot{\boldsymbol{\gamma}}$ and $\boldsymbol{\omega}$ respectively.

The rate of strain tensor is a quantitative measure of the local rate at which neighboring material elements of a deforming continuum move with respect to each other. As a rank-2 symmetric tensor, it has $\frac{n(n+1)}{2}$ independent components which are six in a 3D space. The rate of strain tensor is related to the infinitesimal strain tensor (refer to § 7.3.1) by:

$$\mathbf{S} = \frac{\partial \boldsymbol{\gamma}}{\partial t} \tag{611}$$

where t is time. Hence, the rate of strain tensor is normally denoted by $\dot{\boldsymbol{\gamma}}$ where the dot represents the temporal rate of change, as indicated above.

7.3.8 Vorticity Tensor

This rank-2 tensor is the anti-symmetric part of the velocity gradient tensor and hence it is given by:

$$\bar{\mathbf{S}} = \frac{\nabla \mathbf{v} - (\nabla \mathbf{v})^T}{2} \tag{612}$$

which, in tensor notation with Cartesian coordinates, becomes (see Eq. 100):

$$\bar{S}_{ij} = \frac{\partial_i v_j - \partial_j v_i}{2} \tag{613}$$

The vorticity tensor quantifies the local rate of rotation of a deforming continuum medium. As a rank-2 anti-symmetric tensor, it has $\frac{n(n-1)}{2}$ independent non-zero components which are three in a 3D space. These three components, added to the aforementioned six components of the rate of strain tensor (see § 7.3.7), give nine independent components which is the total number of components of their parent tensor $\nabla \mathbf{v}$.

7.4 Exercises and Revision

7.1 Summarize the reasons for the popularity of tensor calculus techniques in mathematical, scientific and engineering applications.

7.2 State, in tensor language, the definition of the following mathematical concepts assuming Cartesian coordinates of a 3D space: trace of matrix, determinant of matrix, inverse of matrix, multiplication of two compatible square matrices, dot product of two vectors, cross product of two vectors, scalar triple product of three vectors and vector triple product of three vectors.

7.3 From the tensor definition of $\mathbf{A} \times (\mathbf{B} \times \mathbf{C})$, obtain the tensor definition of $(\mathbf{A} \times \mathbf{B}) \times \mathbf{C}$.

7.4 We have the following tensors in orthonormal Cartesian coordinates of a 3D space:

$$\mathbf{A} = (22, 3\pi, 6.3) \qquad \mathbf{B} = (3e, 1.8, 4.9) \qquad \mathbf{C} = (47, 5e, 3.5)$$

$$\mathbf{D} = \begin{bmatrix} \pi & 3 \\ 4 & e \end{bmatrix} \qquad \mathbf{E} = \begin{bmatrix} 3 & e^2 \\ \pi^3 & 7 \end{bmatrix}$$

Use the tensor expressions for the relevant mathematical concepts with systematic substitution of the indices values to find the following: $\text{tr}(\mathbf{D})$, $\det(\mathbf{E})$, $\mathbf{D}^{-1}$, $\mathbf{E} \cdot \mathbf{D}$, $\mathbf{A} \cdot \mathbf{C}$, $\mathbf{C} \times \mathbf{B}$, $\mathbf{C} \cdot (\mathbf{A} \times \mathbf{B})$ and $\mathbf{B} \times (\mathbf{C} \times \mathbf{A})$.

7.5 State the matrix and tensor definitions of the main three independent scalar invariants (I, II and III) of rank-2 tensors.

7.6 Express the main three independent scalar invariants (I, II and III) of rank-2 tensors in terms of the three subsidiary scalar invariants (I_1, I_2 and I_3).

7.7 Referring to question 7.4, find the three scalar invariants (I, II and III) of $\mathbf{D}$ and the three scalar invariants (I_1, I_2 and I_3) of $\mathbf{E}$ using the tensor definitions of these invariants with systematic index substitution.

7.8 State the following vector identities in tensor notation:

$$\begin{aligned}
\nabla \times \mathbf{r} &= 0 \\
\nabla \cdot (f\mathbf{A}) &= f\nabla \cdot \mathbf{A} + \mathbf{A} \cdot \nabla f \\
\mathbf{A} \times (\nabla \times \mathbf{B}) &= (\nabla \mathbf{B}) \cdot \mathbf{A} - \mathbf{A} \cdot \nabla \mathbf{B} \\
\nabla \times (\mathbf{A} \times \mathbf{B}) &= (\mathbf{B} \cdot \nabla) \mathbf{A} + (\nabla \cdot \mathbf{B}) \mathbf{A} - (\nabla \cdot \mathbf{A}) \mathbf{B} - (\mathbf{A} \cdot \nabla) \mathbf{B}
\end{aligned}$$

7.4 Exercises and Revision

7.9 State the divergence and Stokes theorems for a vector field in Cartesian coordinates using vector and tensor notations. Also, define all the symbols involved.

7.10 Prove the following vector identities using tensor notation and techniques with full justification of each step:

$$\nabla \cdot \mathbf{r} = n$$
$$\nabla \cdot (\nabla \times \mathbf{A}) = 0$$
$$\mathbf{A} \cdot (\mathbf{B} \times \mathbf{C}) = \mathbf{C} \cdot (\mathbf{A} \times \mathbf{B})$$
$$\nabla \times (\nabla \times \mathbf{A}) = \nabla (\nabla \cdot \mathbf{A}) - \nabla^2 \mathbf{A}$$

7.11 What is the type, in the form of (m, n, w), of the Riemann-Christoffel curvature tensor of the first and second kinds?

7.12 What are the other names used to label the Riemann-Christoffel curvature tensor of the first and second kinds?

7.13 What is the importance of the Riemann-Christoffel curvature tensor with regard to characterizing the space as flat or curved?

7.14 State the mathematical definition of the Riemann-Christoffel curvature tensor of either kinds in determinantal form.

7.15 How can we obtain the Riemann-Christoffel curvature tensor of the first kind from the second kind and vice versa?

7.16 Using the definition of the second order mixed covariant derivative of a vector field (see § 5.2) and the definition of the mixed Riemann-Christoffel curvature tensor, verify the following equation: $A_{j;kl} - A_{j;lk} = R^i{}_{jkl} A_i$. Repeat the question with the equation: $A^j{}_{;kl} - A^j{}_{;lk} = R^j{}_{ilk} A^i$.

7.17 Based on the equations in question 7.16, what is the necessary and sufficient condition for the covariant differential operators to become commutative?

7.18 State, mathematically, the anti-symmetric and block symmetric properties of the Riemann-Christoffel curvature tensor of the first kind in its four indices.

7.19 Based on the two anti-symmetric properties of the covariant Riemann-Christoffel curvature tensor, list all the forms of the components of the tensor that are identically zero (e.g. R_{iijk}).

7.20 Verify the block symmetric property and the two anti-symmetric properties of the covariant Riemann-Christoffel curvature tensor using its definition.

7.21 Repeat question 7.20 for the anti-symmetric property of the mixed Riemann-Christoffel curvature tensor in its last two indices.

7.22 Based on the block symmetric and anti-symmetric properties of the covariant Riemann-Christoffel curvature tensor, find (with full justification) the number of distinct non-vanishing entries of the three main types of this tensor (see Eqs. 569-571). Hence, find the total number of the independent non-zero components of this tensor.

7.23 Use the formulae found in question 7.22 and other formulae given in the text to find the number of all components, the number of non-zero components, the number of zero components and the number of independent non-zero components of the covariant Riemann-Christoffel curvature tensor in 2D, 3D and 4D spaces.

7.24 Prove the following identity with full justification of each step of your proof: $R^a_{akl} = 0$.

7.25 Make a list of all the main properties of the Riemann-Christoffel curvature tensor (i.e. rank, type, symmetry, etc.).

7.26 Prove the following identity using the Bianchi identities: $R_{ijkl;s} + R_{iljk;s} = R_{iksl;j} + R_{ikjs;l}$.

7.27 Write the first Bianchi identity in its first and second kinds.

7.28 Verify the following form of the first Bianchi identity using the mathematical definition of the Riemann-Christoffel curvature tensor: $R_{ijkl} + R_{kijl} + R_{jkil} = 0$.

7.29 What is the pattern of the indices in the second Bianchi identity?

7.30 Write the determinantal form of the Ricci curvature tensor of the first kind.

7.31 Starting from the determinantal form of the Ricci curvature tensor of the first kind, obtain the following form of the Ricci curvature tensor with justification of each step in your derivation: $R_{ij} = \partial_j \partial_i \left(\ln \sqrt{g}\right) + \Gamma^a_{bj} \Gamma^b_{ia} - \frac{1}{\sqrt{g}} \partial_a \left(\sqrt{g} \Gamma^a_{ij}\right)$.

7.32 Verify the symmetry of the Ricci tensor of the first kind in its two indices.

7.33 What is the number of distinct entries of the Ricci curvature tensor of the first kind?

7.34 How can we obtain the Ricci curvature scalar from the covariant Riemann-Christoffel curvature tensor? Write an orderly list of all the required steps to do this conversion.

7.35 Make a list of all the main properties of the Ricci curvature tensor (rank, type, symmetry, etc.) and the Ricci curvature scalar.

7.36 Outline the importance of the Ricci curvature tensor and the Ricci curvature scalar in characterizing the space.

7.37 Write, in tensor notation, the mathematical expressions of the following tensors in Cartesian coordinates defining all the symbols involved: infinitesimal strain tensor, stress tensor, first and second displacement gradient tensors, Finger strain tensor, Cauchy strain tensor, velocity gradient tensor, rate of strain tensor and vorticity tensor.

7.38 Which of the tensors in question 7.37 are symmetric, anti-symmetric or neither?

7.39 Which of the tensors in question 7.37 are inverses of each other?

7.40 Which of the tensors in question 7.37 are derived from other tensors in that list?

7.41 What is the relation between the first and second displacement gradient tensors?

7.42 What is the relation between the velocity gradient tensor and the rate of strain tensor?

7.43 What is the relation between the velocity gradient tensor and the vorticity tensor?

7.44 What is the relation between the rate of strain tensor and the infinitesimal strain tensor?

7.45 What are the other names given to the following tensors: stress tensor, deformation gradient tensors, left Cauchy-Green deformation tensor, Cauchy strain tensor and rate of deformation tensor?

7.46 What is the physical significance of the following tensors: infinitesimal strain tensor, stress tensor, first and second displacement gradient tensors, Finger strain tensor, velocity gradient tensor, rate of strain tensor and vorticity tensor?

References

G.B. Arfken; H.J. Weber; F.E. Harris. *Mathematical Methods for Physicists A Comprehensive Guide*. Elsevier Academic Press, seventh edition, 2013.

R.B. Bird; R.C. Armstrong; O. Hassager. *Dynamics of Polymeric Liquids*, volume 1. John Wiley & Sons, second edition, 1987.

R.B. Bird; W.E. Stewart; E.N. Lightfoot. *Transport Phenomena*. John Wiley & Sons, second edition, 2002.

M.L. Boas. *Mathematical Methods in the Physical Sciences*. John Wiley & Sons Inc., third edition, 2006.

J. Bonet; R.D. Wood. *Nonlinear Continuum Mechanics for Finite Element Analysis*. Cambridge University Press, first edition, 1997.

C.F. Chan Man Fong; D. De Kee; P.N. Kaloni. *Advanced Mathematics for Engineering and Science*. World Scientific Publishing Co. Pte. Ltd., first edition, 2003.

T.L. Chow. *Mathematical Methods for Physicists: A concise introduction*. Cambridge University Press, first edition, 2003.

P. Grinfeld. *Introduction to Tensor Analysis and the Calculus of Moving Surfaces*. Springer, first edition, 2013.

J.H. Heinbockel. *Introduction to Tensor Calculus and Continuum Mechanics*. 1996.

D.C. Kay. *Schaum's Outline of Theory and Problems of Tensor Calculus*. McGraw-Hill, first edition, 1988.

K.F. Riley; M.P. Hobson; S.J. Bence. *Mathematical Methods for Physics and Engineering*. Cambridge University Press, third edition, 2006.

T. Sochi. *Tensor Calculus Made Simple*. CreateSpace, first edition, 2016.

T. Sochi. *Introduction to Differential Geometry of Space Curves and Surfaces*. CreateSpace, first edition, 2017.

I.S. Sokolnikoff. *Tensor Analysis Theory and Applications*. John Wiley & Sons, Inc., first edition, 1951.

B. Spain. *Tensor Calculus: A Concise Course*. Dover Publications, third edition, 2003.

J.L. Synge; A. Schild. *Tensor Calculus*. Dover Publications, 1978.

D. Zwillinger, editor. *CRC Standard Mathematical Tables and Formulae*. CRC Press, 32nd edition, 2012.

Index

Absolute
    covariant derivative, 130
    derivative, 6, 11, 94, 109, 129–133, 136
    differentiation, 94, 112, 129–132, 136
    permutation tensor, 8, 81, 90, 99, 100, 107, 110
    tensor, 57, 58, 72, 76, 81, 88, 90, 107, 108, 167, 172
    value, 11, 80
Active transformation, 33
Addition of tensors, 14, 62, 63, 73
Admissible transformation, 26, 31, 32, 34, 41, 47–49, 62, 112, 141
Affine
    coordinate system, 25, 30, 118, 167
    tensor, 62, 112, 120, 132, 133
    transformation, 30
Algebraic
    addition, 34, 73
    subtraction, 73
Anisotropic tensor, 58, 72
Anti-
    symmetric, 50, 59–62, 72, 73, 77, 78, 81, 83, 84, 107, 169–171, 176–179
    symmetric tensor, 59–62, 73, 177
    symmetry, 59, 61, 73, 107, 169
Area, 6, 39, 76, 103–105, 110, 160
Associate
    metric tensor, 91
    tensor, 94, 109
Associative, 63–65, 73
Associativity, 32, 63, 65
Asymmetric tensor, 59, 174
Axial tensor, 55

Basis
    tensor, 54, 71–73, 141
    vector, 6, 7, 9, 15, 17, 24–28, 32, 39–46, 48, 49, 53–55, 69–72, 74, 76, 77, 91–94, 97, 99, 100, 109, 112–114, 120, 121, 123–125, 128–130, 132, 134–136, 140, 141, 146, 148, 150, 152
Bianchi identity, 171, 172, 179
Block
    symmetric, 170, 178
    symmetry, 169
Bound index, 14, 20

Cartesian, 1, 6, 7, 10, 15, 16, 22–30, 33–36, 39, 42, 47, 48, 51, 53–56, 66, 68, 69, 71, 72, 74, 83, 91–93, 95, 97, 99, 108–111, 113, 118, 119, 121, 134, 137, 138, 140, 146, 152, 155, 161, 173–179
Cauchy
    -Green deformation tensor, 175, 179
    strain tensor, 6, 175, 179
    stress law, 174
    stress tensor, 174
Chain rule, 130, 131
Christoffel symbol, 6, 7, 94, 112–123, 128, 129, 132–135, 144, 153, 171, 173
Circle, 25, 26, 35, 36
Closure, 32
Comma notation, 6, 10, 11, 19, 117
Commutative, 11, 32, 48, 63–65, 67, 73, 74, 85, 92, 124, 131, 135, 168, 178
Commutativity, 63, 84, 93, 126, 163–167
Complex, 9, 11
Composition of transformations, 32, 47
Conjugate
    metric tensor, 91
    tensor, 94, 109
Conserved, 77, 78, 81, 107
Continuity condition, 11, 20, 31, 47, 124, 162, 169, 171, 173
Continuous, 11, 19, 22, 30, 31, 34, 127
Continuum mechanics, 69, 155, 173–175
Contracted epsilon identity, 87
Contraction, 14, 57, 62, 66–69, 73, 74, 82, 96, 124, 126, 132, 135, 137, 139, 144, 152
Contravariant, 7
    basis vector, 6, 24–26, 39–41, 43–45, 48, 49, 53–55, 69–71, 74, 91, 92, 109, 113, 132, 140
    component, 43, 48, 49, 54, 69, 71, 97, 100, 102, 121, 144
    differentiation, 121, 134
    index, 51, 52, 58, 61, 93, 120, 122, 142–144, 168, 170, 172
    Kronecker delta, 132
    metric tensor, 7, 39, 43, 52, 70, 92, 93, 95, 96, 109, 110, 114, 124, 125, 132, 142
    partial derivative, 11
    permutation tensor, 8, 83, 84, 100, 107, 132
    tensor, 32, 50–55, 65, 99, 121, 143, 153
    tensor derivative, 11

181

Coordinate
- curve, 24–27, 35–40, 43–48, 53, 55, 103–106, 123
- surface, 24–26, 35–40, 43–46, 48, 53, 55, 105, 110, 111
- system, 7, 10, 12, 13, 15, 16, 20, 22–36, 39–41, 44–51, 53, 55, 57, 58, 62, 70–72, 74–77, 83, 91–95, 97, 99, 102, 103, 105, 108–113, 118–121, 123–125, 128, 129, 132–138, 140, 144–148, 150, 152, 158, 161, 167, 171, 176

Covariant
- basis vector, 6, 24–27, 39–41, 43–45, 48, 49, 53–55, 69–71, 74, 91, 92, 99, 109, 113, 114, 132
- component, 43, 48, 49, 54, 69, 71, 97, 100, 101, 121, 141, 143
- derivative, 6, 11, 120–131, 133–136, 141–143, 145, 168, 178
- differentiation, 10, 112, 120–132, 134–136
- index, 51–53, 58, 93, 122, 137, 152, 170, 172, 173
- Kronecker delta, 132
- metric tensor, 7, 39, 42, 43, 52, 53, 70, 81, 91, 93–96, 99, 103, 105, 108–110, 114, 124, 125, 132, 142, 172
- permutation tensor, 8, 99, 132
- Riemann-Christoffel curvature tensor, 167, 169, 171, 178
- tensor, 32, 50–55, 64, 65, 67, 99, 110, 121, 145

Cross product, 43, 56, 57, 84, 99, 100, 103, 104, 110, 137, 139, 143, 152, 157, 163, 164, 177

Curl, 6, 56, 72, 139, 143, 144, 147, 149, 151–154

Curvature
- invariant, 173
- scalar, 173

Curve, 6, 7, 11, 12, 23–25, 35, 36, 53, 76, 98, 99, 103, 110, 129–132, 136

Curved space, 22–24, 46, 94, 109, 110, 178

Curvilinear coordinate system, 24–26, 30, 34, 36, 40, 47, 48, 71, 120, 121, 123, 128, 133, 134, 137

Cylindrical coordinate system, 6, 8, 24–26, 28, 35, 37, 39, 47, 48, 71, 95, 110, 111, 113, 117–119, 134, 137, 148, 150, 152, 153, 176

Determinant, 6, 7, 11, 29, 30, 42, 43, 53, 72, 81, 85–90, 92, 94, 99, 102, 105, 108–110, 115, 126, 142, 156, 158, 172, 177

Diagonal matrix, 44, 95

Differentiable, 31, 121–124, 127, 129–131, 134–136, 138–144, 147, 149–152, 158, 160, 161

Differential
- calculus, 11
- geometry, 12, 17, 24, 119, 155
- operator, 6, 10, 11, 51, 52, 121, 124, 126, 135, 137, 138, 140, 146, 152, 168, 178

Dimension of space, 6, 12, 15–18, 20, 24, 29, 30, 39, 44, 46, 58, 64, 65, 72, 76, 77, 82, 107, 119, 156, 158, 174

Direct
- multiplication, 64–66
- notation, 12
- product, 56–58, 64, 65, 73
- transformation, 29

Displacement
- gradient tensor, 6, 7, 174, 175, 179
- vector, 6, 50, 102–106, 173

Distributive, 18, 64, 65, 67, 73

Distributivity, 163–165, 167

Divergence, 6, 137, 139, 142–145, 147, 149, 151–154
- theorem, 160, 178

Dot product, 27, 42, 43, 53, 66–68, 74, 92, 93, 97, 98, 110, 114, 129, 130, 139, 156, 177

Double inner product, 68, 83, 84

Dummy index, 9, 14–18, 20, 52, 169

Dyad, 6, 55, 56, 65, 68, 72, 83, 122, 148–151

Dyadic
- multiplication, 64
- product, 54

Ellipsoid, 23, 94

epsilon-delta identity, 87, 90, 108

Euclidean, 22–24, 27, 46, 94, 134, 167, 169

Exterior multiplication, 64

Extra pressure, 174

Finger strain tensor, 6, 175, 179

Flat
- metric, 93
- space, 23, 27, 31, 46, 47, 93, 95, 109, 110, 127, 167, 168, 178

Fluid
- dynamics, 138, 173–175
- mechanics, 69, 155, 173, 175

Form-invariance, 77

Form-invariant, 12, 13

Free index, 9, 14–18, 20, 51, 52, 59, 63, 66, 123, 161–167

Gauss theorem, 155

General
- coordinate system, 1, 7, 10, 11, 20, 22, 24, 27, 39, 41–43, 53, 57, 66, 70, 72–74, 90–92, 97, 99, 100, 102, 103, 105, 106, 108–112, 119,

120, 122, 133, 136, 137, 140, 141, 144–146, 148, 150, 152, 153, 161, 167
    tensor, 62, 112, 120, 126, 133, 136, 143
Generalized Kronecker delta, 7, 76, 82, 88–90, 108, 126
Gibbs notation, 12
Gradient, 6, 32, 39, 43, 45, 50, 129, 136–142, 144, 145, 147, 149, 150, 152–154, 175, 176
    vector, 32, 43

Handedness, 32, 55, 81, 105
Homogeneous coordinate system, 23, 27, 29, 47, 110
Hydrostatic pressure, 174

Identity, 32
    tensor, 22, 43, 76, 82, 91, 94
    transformation, 34, 48
Imaginary, 9, 11, 27, 29, 92, 95
Improper
    rotation, 58
    transformation, 32–34, 48, 55, 77, 78, 81, 107
Index
    -free notation, 12, 13
    notation, 9
Indicial
    notation, 12, 13, 17, 19, 126, 135
    structure, 17, 18, 20, 51, 52, 58, 63, 64
Infinitesimal, 6, 22, 23, 34, 46, 91, 103–106, 108, 111, 173
    strain tensor, 7, 173, 175, 176, 179
Inner
    multiplication, 73, 74, 94, 113
    product, 6, 57, 58, 64–69, 74, 83, 84, 94, 97, 113, 124, 126, 127, 129, 131, 156
Integral, 105, 110
    theorem, 155, 160, 161
Intrinsic derivative, 11, 130, 136
Invariance, 12, 16, 19, 71–73, 90, 112, 120
Invariant, 12, 13, 31, 32, 41, 47, 55, 58, 61, 62, 73, 76, 77, 97, 103, 121, 139, 141, 153, 155, 157, 158, 177
Inverse, 32
    Jacobian, 7, 30, 31
    of matrix, 44, 95, 156, 177
    of metric, 44, 91, 92, 95, 109
    of tensor, 175, 179
    transformation, 29–31, 47
Invertible
    matrix, 43, 90, 95, 158
    transformation, 30, 31
Irrotational, 160
Isotropic, 23

tensor, 58, 72, 76–78, 81
Jacobian, 7, 29–33, 39, 42, 47, 48, 57, 94, 105
    matrix, 7, 29–31, 33, 42, 47

Kronecker delta, 7, 15, 41, 53, 58, 71, 76, 77, 81, 82, 84–90, 92, 93, 96, 107–109, 126, 127, 132, 135, 136

Laplacian, 6, 10, 137, 138, 140, 144–146, 148–154, 164
Length, 6, 7, 15, 22, 23, 25, 27, 29, 39, 46, 53, 76, 83, 91, 92, 94, 95, 102, 103, 108, 110, 157
Levi-Civita
    identity, 87
    tensor, 58, 77
Linear
    algebra, 42, 44, 67, 74, 95, 155, 156
    combination, 113
    coordinate system, 25
    operation, 124
    transformation, 26, 27, 29, 34, 62, 112, 121
Lorentz transformations, 23, 29, 95, 110
Lowering operator, 52, 71, 93, 94, 97, 112–114, 126, 168, 171–173

Manifold, 22–24, 31, 39, 42, 46, 93
Matrix
    algebra, 65, 66, 74, 155, 156
    notation, 156
Metric
    space, 22
    tensor, 7, 15, 22, 23, 26, 27, 32, 39, 40, 42–49, 52, 53, 69–71, 76, 81, 90–97, 99, 103, 105, 107–110, 112–114, 118, 119, 123–126, 129, 131–136, 142, 147, 148, 150, 167–169, 171, 172
Minkowski
    metric, 95, 110
    space, 23, 29
Mixed
    derivative, 10, 11, 127, 128, 135, 168, 169, 178
    Kronecker delta, 7, 77, 82, 92, 132
    metric tensor, 7, 22, 41, 43, 91, 92, 109, 124, 125, 132
    Riemann-Christoffel curvature tensor, 168, 169, 178
    tensor, 7, 14, 18–22, 32, 41, 43, 51, 52, 54, 63, 65, 66, 69, 71, 77, 81, 82, 91–93, 102, 109, 121, 122, 124, 125, 131, 132, 141, 143, 152, 168, 169, 178
Multiplication
    by scalar, 62, 64, 73

of matrices, 67, 73, 92, 156, 177
of matrix by vector, 67, 156
of tensors, 14, 58, 62, 64–67, 73, 74, 88, 94, 113, 124, 131

Mutually
  exclusive, 50
  independent, 24, 40, 44
  orthogonal, 15, 25, 26, 36, 39, 43, 44, 49, 53, 83, 92
  perpendicular, 24–27, 55, 94

nabla operator, 6, 137–140, 143, 146, 148, 150, 152, 153
Negative orthogonal transformation, 34, 48
Non-
  Euclidean, 167
  scalar tensor, 24, 32, 64, 65, 112, 120, 121, 128, 133, 134, 136, 146
  singular, 90, 92
Nonlinear transformation, 26, 34
Normal
  stress, 174
  vector, 7, 174
Normalized vector, 6, 102

Oblique coordinate system, 25, 26
Order
  of derivative, 10, 11, 127, 128, 135, 168, 169, 178
  of indices, 10, 16–21, 51, 52, 55, 64, 66, 83, 93, 109, 119, 123, 134, 141, 162
  of multiplicands, 67, 73, 124, 131, 157, 162–164
  of operators, 11, 18, 138, 162
  of tensor, 14, 16, 18, 20
  of transformations, 32
Orthogonal, 15, 25–27, 36, 39, 43, 44, 49, 53, 83, 92
  coordinate system, 7, 24–26, 36, 39, 41, 43, 47, 70, 71, 74, 94, 95, 103–105, 108–111, 115–117, 119, 133, 134, 137, 146–148, 150, 153
  transformation, 27, 33, 34, 48, 55, 78, 81
Orthonormal
  basis set, 15, 43, 49, 53, 71, 83, 148, 150
  Cartesian, 15, 16, 22, 23, 27, 28, 30, 39, 48, 51, 55, 66, 68, 71, 74, 83, 91–93, 95, 97, 99, 108–111, 118, 119, 134, 177
  coordinate system, 15, 16, 20, 22, 23, 27, 28, 30, 39, 48, 51, 55, 66, 68, 71, 74, 83, 91–93, 95, 97, 99, 108–111, 118, 119, 134, 177
  vectors, 6, 15, 20, 49, 72, 83, 84, 99, 108
Orthonormalized vectors, 6, 53, 72
Outer
  multiplication, 64, 67, 73
  product, 6, 64–68, 94, 124, 126, 131, 138

Parallelepiped, 41, 105
Partial
  derivative, 6, 10, 11, 30, 31, 53, 95, 96, 112–114, 117, 118, 120–124, 126–130, 132, 133, 135, 168, 169
  differential operator, 11, 51, 52, 121, 138
  differentiation, 10, 82, 96, 120, 121, 124, 134, 135
Passive transformation, 33
Perimeter, 161
Permissible transformation, 31, 62, 94, 97
Permutation
  of tensor, 68, 69, 74
  tensor, 8, 57, 76–78, 80–90, 99, 100, 102, 107, 110, 132, 139
Perpendicular, 6, 24–27, 39, 53, 55, 94
Physical
  basis vector, 71, 74
  component, 69–71, 75, 137, 147, 148, 150, 153
  dimension, 25, 46, 69
  representation, 6, 39, 69, 71, 74
Plane, 23, 25, 26, 35, 36, 46, 55, 93
Polar
  angle, 36
  coordinate system, 8, 36, 148
  tensor, 55
Polyad, 55
Position vector, 6, 7, 39, 53, 102, 146, 158, 161
Positive orthogonal transformation, 33, 34, 48
Principle of invariance, 12, 19, 112
Product rule, 96, 114, 121–125, 130, 132, 141, 142, 145, 162, 163, 165, 166
Proper
  rotation, 58
  tensor, 55
  transformation, 32–34, 48, 77, 78, 81, 107
Pseudo
  tensor, 55–58, 63, 72, 78, 81
  vector, 55, 57, 72, 157

Quotient rule of
  differentiation, 69
  tensor, 62, 69, 74

Radius, 23
Raising operator, 52, 71, 93, 94, 97, 112, 121, 126, 141, 142, 153, 168, 171–173
Range of index, 12–16, 18, 65, 79, 80, 160
Rank
  -0 tensor, 9, 12, 13, 16, 19, 58, 62, 65, 67, 153, 157

-1 tensor, 9, 12–14, 19, 50, 56, 58, 64, 67, 70, 78, 140, 152, 157

-2 tensor, 9, 10, 12–14, 16, 19, 22, 54, 56, 58–60, 64–68, 70–74, 76, 78, 79, 84–86, 90, 91, 97, 102, 107, 110, 121, 122, 134, 136–139, 141, 143, 149, 151–153, 156–158, 160, 161, 172–177

-3 tensor, 12–14, 19, 54, 60, 66, 72, 77–81, 83, 84, 86, 87, 99, 107, 112

-4 tensor, 65, 66, 72, 79, 107, 167

of tensor, 9, 12, 14, 16, 19, 20, 42, 50–54, 58, 59, 61–70, 72, 73, 77, 78, 84, 88, 93, 97, 107, 120, 123, 128, 130–132, 135, 137–143, 146, 152, 157, 160, 161

Rate of
  deformation tensor, 176, 179
  strain tensor, 7, 176, 177, 179
Real, 9, 11, 24, 27, 29, 35, 44, 103
Reciprocal, 30, 31, 39, 41, 43, 44, 53, 63, 91, 92, 95
  metric tensor, 91
Reciprocity relation, 53, 54, 72, 91
Rectangular
  Cartesian, 25, 26, 29, 33, 34, 54, 56, 72, 92
  coordinate system, 25, 26, 29, 33, 34, 54, 56, 62, 72, 92, 112, 129, 132, 146
  parallelepiped, 105
Rectilinear coordinate system, 24–26, 30, 34, 35, 47, 121, 123–125, 129, 132, 136
Reference frame, 32
Reflection, 33, 34, 55
Relative
  permutation tensor, 8, 81, 99, 107, 110
  scalar, 94, 128, 136
  tensor, 7, 57, 58, 63, 66, 72, 78, 81, 88, 107, 128, 136
Replacement operator, 52, 82, 87, 93, 107, 126, 136
Rheology, 175
Ricci
  curvature scalar, 7, 172, 173, 179
  curvature tensor, 7, 172, 173, 179
  theorem, 114, 115, 124–126, 135
Riemann-Christoffel curvature tensor, 7, 22, 23, 127, 128, 167–172, 178, 179
Riemannian
  curvature, 23
  geometry, 22, 24, 46
  metric, 22, 94
  space, 22, 23, 46, 93, 170
Right handed system, 41, 42, 49, 84
Rotation, 32–34, 58, 175, 177

Scalar, 6, 9–13, 16–20, 31, 34, 58, 59, 62–66, 68–70, 72, 73, 94, 103, 123, 124, 128–132, 135–145, 147–151, 153, 157
  field, 19, 50, 129, 136, 152–154, 158
  invariant, 155, 157, 158, 177
  multiplication, 34
  operator, 137, 138, 152
  triple product, 41, 84, 100, 101, 105, 106, 110, 111, 157, 177
Scale factor, 7, 36, 39, 41, 48, 69–71, 95, 100, 103, 105, 110, 111, 117–119, 146, 148, 150, 153
Schur theorem, 23, 46
Semi-
  circle, 26, 36
  plane, 26, 35
Semicolon notation, 6, 10, 11, 19, 124
Shear stress, 174
Shifting operator, 40, 52, 93, 97, 109, 113, 114, 126, 135, 136
Skew-symmetric, 59, 73, 169
Solenoidal, 160
Sphere, 23, 25, 26, 35, 36, 46
Spherical coordinate system, 6, 7, 10, 24–26, 28, 35, 38, 39, 47, 48, 71, 75, 95, 110, 111, 113, 117–119, 134, 137, 150, 152–154, 176
Stokes theorem, 155, 160, 161, 178
Straight line, 23, 25, 26, 35, 36, 47, 55
Stress tensor, 8, 174, 179
Subtraction of tensors, 62, 63, 73
Sum rule, 132
Summation, 11, 14–16, 20, 43, 62, 66, 113, 147
  convention, 11, 14, 15, 19, 82, 87, 137, 148, 150
Surface, 6, 7, 12, 23–25, 35, 46, 53, 76, 93, 94, 103–105, 119, 160, 161, 174
Symbolic notation, 9, 12, 19, 65, 67, 124, 126, 131, 135, 139, 156
Symmetric, 18, 22, 50, 59–62, 72, 73, 76, 77, 84, 90, 92, 93, 109, 113, 116, 118, 133, 169–176, 178, 179
  tensor, 59–62, 73, 84, 172, 176
Symmetry, 59, 61, 62, 73, 84, 107, 108, 119, 144, 153, 169, 173, 179

Tangent vector, 32, 39, 43, 55, 98, 103, 129, 131
Tensor
  algebra, 42, 62, 63, 156
  calculus, 1, 11, 12, 14, 17, 19, 22, 24, 57, 62, 65, 67, 69, 75–77, 86, 90, 112, 121, 137, 152, 155, 157, 161, 167, 168, 173, 177
  component, 18, 20, 55, 78, 85, 112, 123, 132, 143, 152
  equality, 14–18, 20, 58, 62, 63, 72, 73, 77

expression, 14–18, 20, 56, 58, 63, 72, 87
field, 19, 21, 136, 152, 160, 161
identity, 76, 87, 89, 155, 161
multiplication, 64, 65, 73, 124, 131
notation, 12, 13, 17, 71, 84, 91, 99, 109, 110, 137, 152, 153, 155, 157, 158, 160, 161, 174–179
representation, 50, 69–71, 74
term, 14–18, 20, 56–58, 60, 62, 63, 72, 82, 84, 86, 120, 122–124, 126, 128, 132, 134–136, 144, 162, 164, 172
test, 69, 74
Torus, 23
Total
    derivative, 112, 130, 132, 133, 136
    differentiation, 112, 130, 132, 136
    stress tensor, 174
Totally
    anti-symmetric, 61, 72, 78, 81, 83, 84, 107
    covariant Riemann-Christoffel curvature tensor, 167
    symmetric, 61, 72
Trace, 7, 66, 82, 139, 156, 158, 177
Traction vector, 7, 174
Transformation, 7, 11–13, 15, 19, 22, 26, 27, 29–34, 36, 40–42, 47–51, 55, 57, 58, 61, 62, 69, 71–74, 76–78, 81, 91–94, 97, 103, 105, 107, 109, 112, 121, 141, 153, 157
Translation, 33, 34, 175
Triad, 54, 65
True
    scalar, 31, 157
    tensor, 55–58, 63, 72
    vector, 31, 55, 57, 72

Unit
    tensor, 14
    vector, 55, 56, 84, 160
Unity
    matrix, 53
    tensor, 14, 92, 94, 109, 175

Vector, 6, 7, 9, 11–13, 15–17, 19, 20, 25–27, 31, 32, 39–44, 48, 53–59, 64–74, 83, 84, 87, 91–94, 97–103, 105, 108, 110, 111, 113, 121–124, 127–131, 134, 137–144, 146, 147, 149–153, 156–158, 160, 163, 164, 168, 174, 176, 177
    algebra, 155
    calculus, 74, 137, 140, 147, 148, 152, 155, 157, 158
    field, 19, 138, 152–154, 158, 160, 178
    identity, 87, 161, 177, 178

notation, 12, 158, 160, 178
operator, 137, 138, 140, 152
triple product, 101, 102, 110, 157, 177
Velocity
    gradient tensor, 6, 175, 176, 179
    vector, 7, 175
Volume, 6, 7, 39, 41, 105, 106, 111, 160
Vorticity tensor, 7, 8, 176, 177, 179

Weight of tensor, 7, 53, 57, 58, 63, 64, 66, 72, 73, 78, 81, 88, 94, 107, 128, 136

Zero
    tensor, 14, 58, 62, 72, 73, 133
    vector, 58

Author Notes

- All copyrights of this book are held by the author.
- This book, like any other academic document, is protected by the terms and conditions of the universally recognized intellectual property rights. Hence, any quotation or use of any part of the book should be acknowledged and cited according to the scholarly approved traditions.

Made in the USA
Las Vegas, NV
30 January 2022